FIGHTING FOR THE FUTURE OF FOOD

SOCIAL MOVEMENTS, PROTEST, AND CONTENTION

Series Editor: Bert Klandermans, Free University, Amsterdam

Associate Editors:
Ron R. Aminzade, University of Minnesota
David S. Meyer, University of California, Irvine
Verta A. Taylor, University of California, Santa Barbara

(continued on page 263)

Fighting for the Future of Food

ACTIVISTS VERSUS AGRIBUSINESS IN THE STRUGGLE OVER BIOTECHNOLOGY

Rachel Schurman and William A. Munro

Social Movements, Protest, and Contention 35

University of Minnesota Press
Minneapolis
London

Portions of chapter 3 were previously published as Rachel Schurman and William A. Munro, "Ideas, Thinkers, and Social Networks: The Process of Grievance Construction in the Anti-Genetic Engineering Movement," *Theory and Society* 35 (2006): 1–38; reprinted with kind permission from Springer Science + Business Media. Portions of chapter 4 were previously published as Rachel Schurman and William Munro, "Targeting Capital: A Cultural Economy Approach to Understanding the Efficacy of Two Anti-Genetic Engineering Movements," *American Journal of Sociology* 115, no. 1 (July 2009): 155–202, and as Rachel Schurman, "Fighting Frankenfoods: Industry Structures and the Efficacy of the Anti-Biotech Movement in Western Europe," *Social Problems* 51, no. 2 (May 2004): 243–68.

Lyrics to "You Have No Right" appear courtesy of Tina Bridgman from her album *Hearth;* available on iTunes. Also available are her albums *Deep Sun Inside* and *In the Company of You.*

Published by the University of Minnesota Press
111 Third Avenue South, Suite 290
Minneapolis, MN 55401-2520
http://www.upress.umn.edu

Library of Congress Cataloging-in-Publication Data

Schurman, Rachel.
    Fighting for the future of food : activists versus agribusiness in the struggle over biotechnology / Rachel Schurman and William A. Munro.
        p.   cm. — (Social movements, protest, and contention ; 35)
    Includes bibliographical references and index.
    ISBN 978-0-8166-4761-3 (alk. paper) —
    ISBN 978-0-8166-4762-0 (pbk. : alk. paper)
    1. Food supply.   2. Biotechnology.   3. Agricultural industries.
  I. Munro, William A.   II. Title.
    HD9000.5.S3768 2010
    338.1'7—dc22

                                                                    2010015555

Printed in the United States of America on acid-free paper

The University of Minnesota is an equal-opportunity educator and employer.

18 17 16 15 14 13 12 11 10   10 9 8 7 6 5 4 3 2 1

For Nadia and Eli,
Fiona and Cara

CONTENTS

The Contending Worlds of Biotechnology

On a cool, sunny morning in October 2001, one of us (Rachel) boarded an Amtrak train bound for Washington, D.C., to attend a conference called the Future of Biotechnology. Convened at the luxury Renaissance Washington Hotel, the conference was organized by the Food and Drug Law Institute (FDLI) and cosponsored by three organizations that formally represented the U.S. food and agricultural industries.[1] Among the list of participants were representatives from the food and drug industries, U.S. government agencies, agricultural biotechnology firms, investment banks, the consulting industry, private foundations, and the media.[2]

Even before the lunchtime keynote by Hugh Grant, chief operating officer of Monsanto, the tension in the air was palpable. Bruce Skillings, an industry consultant, opened his presentation with a joke, quipping, "The future ain't what it used to be." He went on to note that although a major report had shown that the current food safety system "provides assurances that foods derived from biotechnology are 'at least as safe' as traditionally derived foods," some groups were still reluctant to accept biotechnology. Larisa Rudenko, director of another Washington-based consulting firm, followed with her concerns about the industry's ability to attract investors. In her view, the industry needed to work much harder to improve public perceptions and consumer acceptance of genetically modified (GM) food, cautioning that if it didn't, "there won't be any more funding [referring to venture capital], and without funding, there *is* no ag biotech industry."[3]

A sense of crisis characterized the day's proceedings. Industry officials pointed to problems they were encountering as they sought to move seed and grain around the world, stemming from the varying trade rules and standards associated with a host of new transnational regulatory regimes for genetically modified organisms (GMOs). A speaker from the seed company Pioneer Hi-Bred accused the European Union of seeking to impose an "impossible tracking from farm to fork of food commodities derived with GM." From the murmurs in the audience, it appeared that most agreed. Kirk Miller, of the North American Grain Export Association, argued that the proliferation of different regulatory requirements was making the grain trade more complex and expensive. While some countries were quite lenient about GMOs, others were setting strict standards. Global grain traders could not easily deal with this sort of market chaos, he intimated.

Speakers expressed particular frustration with the situation in Europe, where they believed prospects had gone from bright to dim because biotechnology had become so politicized on the continent; "unrepresentative" nongovernmental organizations and compliant politicians were wreaking havoc on the industry's plans for the future. Since 1996, activists had succeeded in transforming a generally agnostic attitude toward GM technology into a predominantly resistant one.[4] Well aware of public sentiment, many food processors decided not to use GMOs in their own processed foods, and retailers, including most of the continent's major supermarkets, decided not to sell products that used them as ingredients.

The European Parliament also made an about-face in its treatment of GMOs. Whereas the Parliament had been quite willing to approve new GM crop requests in the mid-1990s, it placed a moratorium on all new GM crop approvals after 1998, even though the decision to do so was clearly in violation of World Trade Organization (WTO) rules. In three short years, the agricultural biotechnology controversy had exploded, and markets for genetically engineered food and seed had effectively closed. "Conditions other than science are affecting the [European] regulations," complained one industry representative, implicitly normalizing what he saw as a pure and unpoliticized "science-based" approach to food regulation and dismissing European politicians' responsiveness to their constituencies' concerns.[5]

At the turn of the twenty-first century, Europe was not the only place where genetic engineering had become subject to rigorous debate. Issues of whether or not agricultural biotechnology would help or hurt farmers, resolve problems of world hunger, harm the environment, or pose threats to public health were disputed in many countries of the global South. In Brazil, the Philippines, and Mexico, government ministries, farmers' unions, consumer organizations, and environmental groups clashed over whether to allow commercial planting of GMOs. In South Africa, new organizations aimed specifically at challenging the introduction of GMOs emerged. In India, a large farmers' movement in the state of Karnataka publicly burned GM field trials and joined with other groups to start the "Cremate Monsanto" campaign, aimed at ending that company's business in GM cotton seed. Thai activists organized a long march through their countryside in September 2000, with the goal of "sensiti[zing] public opinion on the threat of GMOs and the promise of people's alternatives for food security and agricultural biodiversity in Asia."[6]

Even in the United States, in many senses the home of these new technologies, genetic engineering became the subject of social protest and mobilization. In July 2000, *Time* magazine captured the political spirit of the moment:

> It was the sort of kitschy street theater you expect in a city like San Francisco. A gaggle of protesters in front of a grocery store, some dressed as monarch butterflies, others as Frankenstein's monster. . . . People in white biohazard jumpsuits pitching Campbell's soup and Kellogg's cornflakes into a mock toxic-waste bin. . . .
>
> But just as the California activists were revving up last week, similar rants and chants were reverberating in such unlikely places as Grand Forks, N.D., Augusta, Maine, and Miami—19 U.S. cities in all. This was no frolicking radical fringe but a carefully coordinated start of a nationwide campaign to force the premarket safety testing and labeling of . . . genetically engineered organisms. Seven organizations . . . were launching the Genetically Engineered Food Alert, a million-dollar, multiyear organizing effort to pressure Congress, the Food and Drug Administration and individual companies, one at a time, starting with Campbell's soup.[7]

No wonder the industry felt things had spiraled out of control.

It hadn't always been so. When molecular biologists, geneticists, and plant biochemists first developed the ability to cut and splice genes from one organism into another in the 1970s and 1980s, the prospects for this revolutionary new technology looked remarkably open and bright. The scientific profession, the media, venture capital, and Wall Street were abuzz with the possibilities these new "recombinant DNA" technologies held out for generating a whole new industrial frontier and for solving a host of agriculture- and health-related problems. For these enthusiasts, the new biotechnologies offered a novel way to shortcut the slow process of traditional plant and animal breeding, to raise agricultural productivity, and to make better and cheaper medicines, all while representing a potentially enormous source of profit for the firms involved. In the tremendous excitement of this first stage, the biotech scientist–entrepreneurs saw their work as an enterprise in which "everybody wins."

Their enthusiasm was infectious. Large corporations and finance capitalists poured money into these new ventures and built a massive scientific-cum-business infrastructure dedicated to generating new discoveries and new products with recombinant DNA. Indeed, when genetically engineered crops were introduced into the market in 1996, they took off at a phenomenal rate. The first crops planted commercially were corn, soybeans, canola, and cotton. By the time of the FDLI conference, global plantings of these crops had grown from 4.2 million acres in six countries to 130 million acres in thirteen countries, a thirtyfold increase.[8] Some observers hailed this as the most rapid uptake of a new technology in human history.[9] For many, the "gene revolution" of transgenic technology would underpin a second Green Revolution to resolve the challenge of global hunger.[10]

In the light of this powerful sense of possibility, the jitters demonstrated by the participants in the FDLI conference and the vigorous expressions of opposition from some groups around the globe seem anomalous. How are we to make sense of these profoundly divergent responses to the development of genetic engineering technologies: exuberance versus wariness; promotion versus rejection; visions of promise versus the dread of disaster? How did these responses develop, and how did they shape the trajectory of the technology? As we recount in the following chapters, the growing social controversies that so disturbed industry representatives at the FDLI conference were driven at their core by a group of social activists who saw in biotechnology not its infinite promise but a number of profoundly unsettling questions. What would these organisms do once released into the environment? How would the ingestion of genetically modified foods affect human health? Did the technologies harbor dangers of a new eugenics in which individuals could be designed (quite literally) to exhibit particular traits? Would genetic engineering by private companies lead to the commodification of life, and if so, what would that mean for society? Development critics from the global South and North who were concerned about the power of rich countries and multinational corporations had other questions: How would the technology affect the global food system? Would these technologies benefit small-scale farmers or work to their detriment?

From the very moment that scientists discovered how to splice individual genes from one species into another, critics took issue with the new biotechnologies and sought to halt or at least to drastically slow down their development and deployment. Over the next thirty years, these critics devoted themselves to developing a collective analysis of the technology. They educated the public about what they believed were the pitfalls and potential perils of genetic engineering and pushed government policies in a more precautionary direction. Gradually, their criticism and campaigns attracted more and more supporters into what ultimately became a widespread anti-biotechnology movement. In the late 1990s, they expanded their protest repertoire to include the supermarket and corporate campaigns, GM crop sabotage, and creative street demonstrations of the sort described in *Time* magazine.

Wresting agricultural biotechnology into the public sphere, these critics transformed it from an elite technological development into a highly contentious social problem. From the beginning, they insisted that the conversation about biotechnology would not be confined to scientific laboratories, corporate boardrooms, and government offices in the global North but would instead be widely debated within different societies and among different segments of those societies. In broadening the conversation, they challenged the prerogatives of large corporations and the notion that science is, and ought to be, apolitical. Activists led many people to reject GMOs and closed European markets to genetically engineered products, precipitating a market shift that reverberated around the world. They challenged government regulatory policies and used national legal systems and courts to constrain the ability of firms to deploy GM technologies at will. At the global level, they helped to create a multilateral regulatory regime that shaped patterns of GM food trade. By creating alternative discourses at the local and global levels, they countered the industry's and many governments' portrayal of this new technology as an unqualified "social good" and raised issues that pushed many people to think more critically about it.

Anti-biotech activists also imposed large economic costs on biotechnology companies, increasing their risks of investment. Indeed, the industry's early temerity turned into timidity as large firms became much more cautious about developing new applications. At the time of this writing, GMOs remain substantially excluded from Europe and can be sold legally in only two countries on the African continent (South Africa and Burkina Faso). The range of commercially available crops that are based on GM technologies remains largely limited to the initial four: soy, cotton, corn, and canola.

To be sure, the global controversy precipitated by the actions of these activists did not bring about a complete rejection of GMOs. After 2005, in fact, the agricultural biotechnology industry experienced a significant recovery, bolstered by a growing demand for biofuels and the 2008 spike in global food prices. Some small seed companies and farmers also took matters into their own hands, either producing or planting GM seed illegally or both. Nevertheless, anti-biotech activism has had a powerful effect on the development and deployment of the technology, and the use of genetic engineering in agriculture remains politically contentious. By challenging the twin hegemonies of science and profitability as the only meaningful metrics for deciding the direction of new technologies, global activists injected an entirely different set of values into the discussion. Today, all decisions to develop, market, or use transgenic technologies must take these values into account.

The brief history we have just sketched raises some fascinating questions. First, what led some individuals to interpret recombinant DNA technology as a problem in society when the scientific profession, the media, business, and many government policymakers were so enthusiastic about it? This question becomes even more intriguing if we recognize the small scale of the industry when these technologies first appeared. In the 1970s and early 1980s, the science of gene transfer was in its infancy: industry was just starting to become involved with the technology, and experimentation with these techniques was limited mainly to small biotech start-up companies. Research was still being conducted primarily by university-based scientists, and no biotechnology products were on the market. By the time of the FDLI conference in 2001, no regulatory or public health disasters had been associated with GMOs that might have raised a red flag for the consuming public. So what drove some individuals to mobilize in protest against these new technologies? And why did they do so with such intensity and passion?

Moreover, how should we interpret the impact of this movement on the development of the technology? Even though anti-biotech activism has remained alive and present in many countries, the movement is tiny compared with the industry it is confronting. Even at its peak at the turn of the twenty-first century, the movement was small and under-resourced. So how did such a small group of naysayers manage to impose such harm on an economically powerful and politically supported industry? What made the industry so vulnerable to criticism and oppositional mobilization?

Finally, given that GM crops continue to spread around the globe and the agricultural biotechnology industry has gotten back on its feet, to what extent can we say that the movement has been successful and in what ways? These are the questions this book seeks to answer.

Making Biotechnology into a Social Problem

As social movement scholars have long recognized, the mere existence of social inequality, the denial of basic civil rights, or a threat to public health or environmental well-being is not enough to inspire people to take action or even to view any of these situations as a social problem. After all, every society is replete with various forms of injustice, social exclusion, and threats to human health and the environment, but within a given social order not all of these are perceived as unacceptable situations that require remedial action or even as problems capable of being remedied. For a given situation to be widely understood as a social problem (i.e., as a problem that demands attention and remediation), some group has to form an oppositional consciousness around it and identify it as a problem. This requires an *act of interpretation*.[11]

In order to understand the collective acts of interpretation through which anti-biotechnology activists built their movement, we need a sociological and historical analysis of the individuals who first identified genetic engineering as a problem and who attributed to it a very different set of meanings from those embodied in the dominant discourse of problem-solving scientific innovation. In chapter 3, we show in detail how social activists opposing biotechnology expressed a worldview that differed quite significantly from those held by most industry, science, and government proponents of the technology. This countercultural worldview was informed by the particular generation and historical moment in which these activists came of age, as well as by their personal biographies, that is, the events and experiences that shaped their individual lives. Drawing on their experiences in the tumultuous 1960s, being present at the birth of the environmental, feminist, anti-nuclear, and Non-Aligned movements, and observing some of the more pernicious effects of the North's "development project" on the global South, these individuals were disposed to look upon these technological developments with a critical eye. They were also deeply suspicious of the motives of big business. Consequently, they were inclined to assess the technology in its socioeconomic context.

The ideational and normative elements of this countercultural worldview are readily apparent in the interviews we conducted with anti-biotechnology activists. "Really, it's a matter of control and ownership," observed one activist from the United States, "because it's just so clear [that when] you look at what has really happened in the 130 million acres that have been planted and the fact that Monsanto's technology accounts for 91 percent of [that acreage] worldwide, what else do you have to know? We're talking about control by a single company!" Another interviewee, whose concerns were primarily environmental, exclaimed, "[People] don't get that *you can't call it back*. It's not like the blanket consequences of what happens when you spray DDT across a bunch of crops. You know, every single transgenic organism is sui generis, and you don't know what the implications of that organism proliferating in the world are going to be." Within a social context in which others shared similar ideas and the same overall worldview, such views became a call to action. As one activist explained, "Their answer—by 'them,' I mean, the technologists and the corporations—they say, 'No, we're not going to change our technologies so they fit life; we're going to change life so it fits our technology.' That is why I became so involved in biotechnology. That larger image . . . has been so horrific to me."

The act of interpreting the nature and meaning of genetic engineering in such critical terms was grounded in what we might think of as a "cultural predisposition" to question these developments closely. By *cultural predisposition* we refer to the shared mental worlds that incline people toward particular ways of thinking and seeing and that, in this case, have defined certain of the technoscientific developments associated with modernity as problematic.[12] As cultural sociologists have theorized, shared mental worlds involve sets of beliefs, assumptions, images, and value judgments about how the world works (and should work), as well as ways of thinking and categorizing things. Often these are so taken for granted that their bearers are not conscious of them, but they still serve to structure much of human behavior, even while people are not bound to act in accordance with them. However, for us to fully understand how this group transformed its cultural predispositions into a collective and sustained criticism of biotechnology, and from there into social action, our explanation must go beyond social actors' shared mental and moral worlds to include their social circles and intellectual communities. Taken together, these elements constituted the essence of the activists' "lifeworld."

The term *lifeworld* has a long and weighty history and thus requires some attention and definition before we begin using it in this book. Within sociology, the term is usually associated with Alfred Schutz (and his former student Thomas Luckmann) and with Jürgen Habermas.[13] These theorists understood the lifeworld as a preconscious state that is "always already" familiar to people by virtue of having been born into an existing cultural and material world. A lifeworld comprises a stock of culturally transmitted background knowledge that people bring to a situation and that provides them with a common cognitive and normative frame of reference.[14] This "natural attitude," as Schutz called it, is so intuitively familiar to people that they are largely unaware of its existence, even while it mediates every aspect of their everyday experience.[15]

While both Schutz and Habermas developed their concepts of the lifeworld as part of a metatheory of human consciousness and, in Habermas's case, of human communication and societal evolution,[16] our use of the term is much more specific and concrete and emphasizes the way that the local culture of a group (in our case, anti-biotech activists on the one hand and members of the biotechnology industry on the other) and the social world of that group shape the way its members experience, understand, and interpret the world. Lifeworlds, as we conceptualize them, are simultaneously cognitive, moral, and social and are collectively constituted in particular spaces of interaction, or milieus. They come into being through a process of ongoing activity and social interaction among groups of people.[17] As groups of people use and express their ideas, values, and commonsense meanings in regular interactions with friends, colleagues, co-workers, and other associates, they develop certain shared accounts of the world and "normal" ways of acting that are, as Kate Crehan puts it, "so firmly embedded within individuals' consciousnesses as to seem to those individuals part of the very texture of their own subjective being."[18] It is in the process of interacting socially, swapping ideas, and creating shared meanings that people come to consolidate a lifeworld.[19]

The significance of a lifeworld for understanding social action is that it generates, and naturalizes, certain broad visions of the world, as well as interpretations of specific phenomena. These, in turn, predispose people to particular types of behavior. For those who were part of the activist lifeworld, for instance, apprehending genetic engineering in terms of who owned and controlled it, its potentially irreversible impacts on the environment, and the dim picture it painted of modern society and human alienation from the rest of nature made perfect sense. The specific actions in which people engaged also flowed out of their lifeworlds; being part of the activist lifeworld inclined people toward certain norms of behavior and strategies of social action and a priori excluded others. Experience in this and other social movements exposed people to certain repertoires of collective action and inculcated particular ways of thinking about how to produce change, which in turn shaped what activists did on a daily basis. For people in this group, attacking industry's constructions of bioengineering and organizing direct actions against the firms involved in its development were logical courses of action. At least in the United States, working *with* the industry was just as "naturally" ruled out, in that it signified cooperating with the enemy and exposing the activist group to the possibility of political co-optation.

Industry Lifeworlds

It was not only the activists whose worldviews and behavior were powerfully shaped by their local culture, shared values, and work and social circles; the same was also true of industry executives and managers, as well as industry scientists, all of whom were part of an industry lifeworld.[20] Most corporate executives and managers—at least in the main home of the industry, the United States—apprehended biotechnology on two levels. At a concrete level, they thought about genetic engineering in terms of the specific goods it could produce, such as a bovine growth hormone (BGH) that would induce cows to produce more milk or a corn that was herbicide-tolerant, and the economic benefits these goods could generate for the firm. As Richard Mahoney, former CEO of Monsanto, explained to journalist Daniel Charles, "We got into BGH like we got into a lot of things. We'd been making agricultural chemicals for years. You increase the productivity of the farmer; you keep half [the profits] and give him half. So what's the big deal? There wasn't even one discussion of the social implications. I never thought of it."[21]

Ever-present in industry officials' minds as they went about the process of technology development was the need to address the problems of their central customers, namely, midwestern farmers. The words of Mahoney's predecessor at Monsanto, John W. Hanley, clearly reveal such thinking. "One of [our] first aspirations was to develop crops that would permit the random application of [our herbicide] Roundup," he told us. "If we had a corn crop which was Roundup resistant, then the farmer could broadcast that, kill the weeds, not worry about the corn. The corn would grow and take advantage of the increased moisture and nutrients. And everybody would be happy."[22]

At a more abstract level, biotechnology held several other shared meanings that drew upon some of the central tropes and cultural values available in society. To most corporate executives in the biotechnology industry, genetic engineering represented the epitome of scientific progress coupled with economic development. What could be better than engineering plants to produce their own pesticides and boosting a company's business at the same time? It also represented a basic "right," in the sense that corporate managers believed their companies had the right to engage in the scientific research and technology development of their choosing, as long as a potential market existed for those goods and the products they produced were legal.

Both of these specific and general interpretations predisposed corporate managers to engage in certain types of business behavior. Just as the activists' lifeworld drove them to fight patents and organize demonstrations, the lifeworld of industry officials led them to pour resources into technology development and to move as quickly as possible to establish intellectual property rights over their firms' scientific innovations. They also devoted themselves to the task of achieving permissive government regulation, consistent with their belief that restrictive regulation would make it difficult to profit from their investments.

Industry scientists shared key aspects of the corporate worldview common in the industry. Yet they also brought their own culture of science to the workplace. For those studying plant biology, genetics, and biochemistry in the 1970s and 1980s, recombinant DNA represented the cutting edge of their disciplines. This created a tremendous sense of excitement, competition, and drive among these scientists. Industry scientists also possessed a strong belief in the quality of their science and in the validity and robustness of the laboratory methods they used to create and test genetically engineered organisms. Most were convinced that the technology they were developing was more controllable and safer than existing plant-breeding methods.[23] These views were complemented by the perception that the public knew so little about basic science that it was fundamentally ill-equipped to weigh in on complex scientific questions, including the debate over genetic engineering.

Such ideas and beliefs spilled over into the lifeworld of corporate managers, who developed a deep faith in the infallibility of their scientific departments and a notable disdain for those who criticized the technology, especially if those critics were not scientists. In fact, as the industry developed, the lifeworlds of industry scientists and corporate managers increasingly came to overlap. At the same time, both of these industry lifeworlds evolved in contradistinction to the activist lifeworld, particularly after the mid-1990s, when anti-biotech activism began to impinge upon the industry's freedom to commercialize its new technology. This ongoing (and growing) opposition between industry and activist lifeworlds made it difficult for these actors to communicate with and comprehend one another, even though some members of each group ostensibly shared the same values and goals, such as a desire for a more ecologically sustainable agriculture. In the process, each group's lifeworld became increasingly entrenched, albeit in an interactive and dynamic way.

Opportunities for Activism

Students of social movements have long argued that the efficacy of these movements depends not just on what movements do but also on the contexts in which their members carry out their actions, as well as the ongoing interactions between movements and their adversaries. Indeed, the central insight of this perspective is that different social and political contexts provide different political openings for activists. When external conditions are favorable, movements are better able to mobilize and are more likely to achieve their goals than when those conditions are inhospitable.

Most of the scholars who have used this lens to study social movements have focused on organizing opportunities arising out of the domestic political sphere, such as the presence or absence of elite allies in government or the relative openness of the political system.[24] More recent scholarship has indicated the significance of the broader cultural environment in which movements operate, suggesting that culture shapes movement strategies and influence by affecting people's expectations of their countries' political institutions, their sense of their own power, and the resonance of particular discourses in society.[25] Indeed, as we show in chapters 4 through 6, these sorts of political, institutional, and cultural conditions are crucial for understanding the efficacy of the anti-biotechnology movement. At the same time, however, vulnerabilities emanating from the economic sphere also turn out to be of great importance. In the case of agricultural biotechnology, the organization of the global commodity chain into which the industry was inserted and the lifeworlds of industry actors proved to be important arenas of movement opportunities and strategic action.[26] To see why this was so requires that we delve into the particulars of the chain.

Industry Structures and Organization

Figure 1 illustrates a commodity chain for a typical manufactured food product. This commodity chain connects agricultural input providers (a group that came to include agricultural biotechnology companies in the 1990s) to consumers through an extended series of relationships. Several challenges need to be overcome if an agricultural biotechnology company

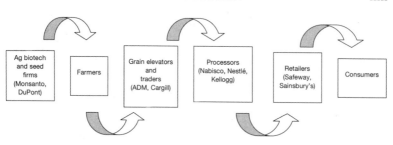

Figure 1. Global commodity chain for typical processed food products.

is to realize the value of its investment in genetically engineered crops: its GM seeds have to be bought and planted by farmers, farmers' crops have to be bought by grain elevators and traders, traders have to sell the grain to food manufacturers to use as ingredients, the output of those food manufacturers has to be purchased by retailers such as supermarkets and commercial food preparers, and finally, these retailers have to sell their products to consumers.

From the perspective of an activist, two features of this chain stand out. First, the agricultural biotechnology industry depends heavily on the other participants in the chain, because it mainly produces products (namely, GM seeds) that do not have many alternative uses.[27] Second, because of the industry's heavy dependence on this particular commodity chain, anti-biotechnology activists can, in theory, exert pressure at any link along the chain to hurt the biotechnology industry economically. More concretely, even if activists can't persuade farmers to stop buying GM seed, they might be able to induce food manufacturers or retailers to stop buying and selling food made with GMOs. Or they can take their case directly to consumers, possibly convincing them not to eat GM food. In short, the commodity chain for food offers activists a number of distinct "choke points" at which they can potentially debilitate the industry.

In practice, of course, the vulnerability of actors located at different choke points varied significantly. The group of actors most vulnerable to anti-biotechnology activism turned out to be supermarkets and food manufacturers located in countries where citizens were extremely sensitive to the safety and quality of their food and to the environmental consequences of producing it. As chapter 4 reveals, the specific sources of these supermarkets' and food manufacturers' weaknesses lay in the high level of competition among the firms composing these two sectors and in what the business literature refers to as a company's "reputational assets," or the status of its store names and brands. By challenging the depth of these companies' commitments to providing safe food to consumers and putting them at risk of losing market share, anti-biotechnology activists, in Europe at least, were able to dissuade these firms from using GM ingredients and from carrying GM food in their stores.

The transnational scope of the commodity chain for many food products carried additional implications for activism and the pressure points at which it could be directed. Most obviously, this broad scope meant that activism against agricultural biotechnology in one part of the world could be used to affect industry fortunes in another. Indeed, as we illustrate in the latter half of this book, the relations of economic dependency that tied geographically dispersed chain participants together guaranteed that the effects of activism would reverberate widely. Somewhat less obviously, this scope meant that in order to sell their products, biotechnology companies had to work through literally dozens of national and regional regulatory agencies whose job it was to regulate the food and agricultural sectors, including the use of new technologies therein. Activists could thus attempt to influence these regulatory systems directly or indirectly if they wanted to interrupt the global spread of GMOs. This turned out to be quite an effective strategy in a number of political contexts.

An additional source of openings available to activists can be traced to the lifeworlds of industry actors. Although social movement scholars have long recognized the significance of adversary behavior in shaping movement strategies and efficacy, they have not widely examined *cultural influences* on that behavior.[28] Especially neglected has been the culture of firms and other economic actors. Recent scholarship in economic anthropology, geography, and history can help address this lacuna by recognizing that economic actors are also *cultural actors* whose understandings of the world, as well as of their own operating environments, are profoundly influenced by culture.[29]

This literature suggests that corporate culture (or in our terminology, the industry's lifeworld)[30] affects the way industry actors read the world as well as how they act upon it. Corporate culture influences, but by no means determines, individual actors' sense of what is normal, reasonable, and "smart" behavior in the context of their business worlds. As Erica Schoenberger suggests, corporate culture is inextricably tied to corporate strategy. More specifically,

> the relationship between corporate culture and strategy is . . . more intimate than is normally conceived. Strategy embodies knowledge and interpretations about the world and the firm's position in it. It is an

exercise in imagining how the world could be or how it ought to be. In this light, strategy is produced by culture. At the same time, since the past strategic trajectory of the firm embodies specific configurations of practices, relations, and ideas, culture is also produced by strategy. The two are mutually constitutive categories.[31]

In other words, corporate culture (or an industry's lifeworld) influences an industry's behavior—both everyday behavior and behavior that is more conscious, purposeful, and specifically oriented toward the future.

For our purposes, the significance of these observations is that industry actors' culturally inflected ways of reading the world and acting upon it can create important political openings for activists. In chapter 4, we illustrate this point with our discussion of Monsanto's deeply misguided approach to introducing GMOs into the western European market. As we show there, Monsanto moved into western Europe like a bull in a china shop, operating in a manner that reflected the parochially competitive lifeworld of its most powerful executives and the company's characteristically aggressive business culture. The inability of these executives to read Europe's political climate and consumer culture accurately, and their decision to ignore the warnings of British allies and some of their own European employees that an aggressive approach would backfire, made the company vulnerable to activist attacks on its reputation.

The corporate culture of specific firms and industries also influences how they respond to challengers. For example, as our discussions of the biotechnology struggles in Europe and the United States reveal, the distinct responses of the biotechnology and food industries to GMO-labeling initiatives in each of these contexts created different organizing opportunities for European and U.S. activists. In Europe, the agricultural biotechnology and food industries accepted the implementation of stringent labeling requirements by the European Union, which made it easier for activists to stigmatize GM products and for consumers to avoid them. In the United States, by contrast, both of these industries successfully fought against activists' labeling campaigns and thus closed off this avenue of consumer mobilization.[32]

Studying the Biotechnology Controversy

The controversy over agricultural biotechnology unfolded on a global level, in large part because the technology itself is global. It is embedded in a globally integrated agrofood system characterized by global seed and commodity markets, multinational corporations, and international trade regimes. The governance of the technology involves laboratory-level re-search protocols, national risk assessment and biosafety regulations, inter-national trade agreements, and transnational rules for the protection of intellectual property rights and global market regulation. How this tech-nology is produced, evaluated, and governed, therefore, takes place at a variety of levels, ranging from the intimacy of the scientific laboratory to the negotiating chambers of the World Trade Organization.

In important respects, the anti-biotech movement can also be under-stood as a transnational movement. From the early days of activism in the 1980s, leading anti-biotech activists interacted and consulted inter-continentally from their home bases around the world. They organized international conferences, shared ideas and information, and supported one another's efforts. The intellectual architecture of the anti-biotech movement was constructed in these interactions, and in that sense the movement was inherently transnational, as we show in chapter 3. In ad-dition, anti-biotech activists networked with allies in other transnational social movements and were an important part of the growing social move-ment against corporate globalization, which burst onto the international scene in Seattle in 1999. The anti-biotech movement also emerged and organized at the level of international regulatory institutions, helping to produce the rules set by those institutions and, thus, new international norms. Two important examples include anti-biotech activists' influence in shaping EU food and biotechnology policy and their role in helping determine the terms of the Cartagena Protocol on Biosafety, an inter-national agreement that governs trade in genetically modified organisms. In this respect, the anti-biotech movement is part of a broader panoply of transnational organizations and movements that operate in what Sidney Tarrow terms a "triangular opportunity space" shaped by international-ism, that is, a space that is structured by relations among states, nonstate actors, and international institutions and that produces opportunities for actors to engage in collective action at different levels.[33]

But the transnational characteristics of the industry, the technology, and the movement should not obscure the fact that most of the cam-paigns carried out by activists have been local and articulated in terms of local concerns and interests. Anti-biotech activists have typically led marches through their own cities and countrysides, organized super-market campaigns at the local or national level, and mobilized discourses that will resonate with the citizens of their own countries. For the most part, they have targeted state authorities, domestic firms, or the local affili-ates of transnational corporations. In this respect, the anti-biotech movement has been as much a collection of local movements as it has been a transnational movement, even though local struggles have been neither independent of one another nor constrained to the places in which they are waged. "Local" anti-biotech movements then, as now, have worked together, inspired one another, shared resources, and participated in one another's struggles.

The political field of contention around agricultural biotechnology has thus been immensely complex. It has involved both global and local struggles that bring together the fate of farmers' livelihoods and the individual choices of consumers on the one hand and the logics of the global food economy on the other. These multilevel struggles have engaged a range of social and professional networks as well as transnational discourses of modernity, development (modernization), and authoritative knowledge. As with any global controversy, the relationships among people, places, situations, and events have been multifaceted, interdependent, and often opaque. Actors are commonly motivated by multiple, and sometimes conflicting, concerns and interests. The degree of fusion between local movements and transnational movements varies widely. Moreover, the specific conflict over agricultural biotechnology cannot be neatly cordoned off from other controversies, since the actors involved in this struggle have invariably drawn on their histories in other contexts. Given this complicated terrain, causality has been neither simple nor singular, and the task of specifying the relative causal weights of local and global actions, discourses, and policy shifts in shaping the development trajectory of a technology poses knotty analytical challenges. In particular, how does one measure the influence of social activism on the ground, as it moves across geographical space and as it crosses different institutional and cultural settings? How does one understand the relationship between different "local" social movements and the transnational movement of which they are a part?

To address these challenges, we adopt two analytical strategies. First, we agree with scholars who argue that, because the nation remains citizens' principal "imagined community" for claiming rights, and the state provides the overriding institutional environment for structuring contention, social movements tend to organize and strategize at the local and national levels.[34] For this reason, our discussion focuses mainly on movements acting in situ, at either the national or the regional scale. Each of the specific conflicts we study (in western Europe, the United States, and Africa) had its own political and cultural dynamics. Yet, in the context of a global industry and transnational networks of actors, these conflicts were also interconnected in ways that affected not only patterns of activism in specific locations but also their larger impacts on the technology's trajectory. For instance, although activism in the United States had a negligible impact on the domestic policy environment, U.S. activists played a crucial role in creating the Cartagena Protocol on Biosafety, which became a major resource for activists in poor countries to pressure their governments. And European activists effectively closed national supermarket chains to GM products, which not only forced biotechnology companies to reconsider the economic calculus of product development but also significantly influenced African governments' approaches to regulating the entry of GMOs into their countries. In order to capture the interaction between local struggles and transnational impacts, we adopt an analytical strategy of "relational comparison," which does not consider local conflicts simply as self-contained case studies of larger processes but instead recognizes that, in the context of a transnational political opportunity space, these conflicts are formed in relation to one another and to a larger whole.[35]

Second, in our view, the best way to deal with the complex causality of a multiscale and multidimensional movement is through a narrative and interpretive style of analysis. Chronicling what happened when and why different actors interpreted the meaning of specific actions and events differently allows us to understand the positionality of different subjects, the interplay among actors, and the historical sequencing of events within a structuring environment. It also leads us to recognize the significance of "path dependency," or the idea that what happens at one moment in time often affects what is possible (or likely to happen) at subsequent moments. In other words, when actors undertake specific actions, these actions often close or partly close some doors of possibility and open up others. This helps explain not only how struggles unfold but also how the process of change lurches forward.

In the following chapters, we bring these two analytical strategies to bear on our study of the biotechnology controversy. In chapter 1 we outline the historical precursors to the controversy in terms of the emergence of knowledge-based economies, the rise of a hegemonic neoliberal ideology, and the growth of new social movements. Against the background of this historical moment, we then examine how different groups of actors interpreted the world through their lifeworlds and how those lifeworlds influenced their behavior. In chapter 2 we focus on the construction of the industry lifeworld, revealing how individual companies charged ahead, enthralled by their scientific discoveries and driven by competitive pressure to establish their positions in the market for agricultural bio-technology. In chapter 3 we map the construction of the activist lifeworld, exploring anti-GMO activists' worldviews and motivations and showing how they formed an oppositional ideology. This chapter illustrates the various cognitive, social, and material processes by which early activists built an intellectual architecture for the movement.

In subsequent chapters, we turn to the task of specifying the respective dynamics of contention in Europe (chapter 4), the United States (chapter 5), and Africa (chapter 6). In each of these chapters, we demonstrate how these struggles influenced the trajectory of the technology both in situ and on a global scale. Consistent with our relational comparison approach, we pay particular attention to the ways in which ideas, information, and repertoires of action flowed among these locations, reflecting the relationship between national and transnational activism. In the Conclusion, we summarize our argument about how the anti-biotech movement mattered and speculate on the future of agricultural biotechnology. We end with a brief discussion of how our book contributes to contemporary thinking about social movements.

A Tale of Two Lifeworlds

In May 2003, I, Rachel, drove westward through Illinois to attend two back-to-back conferences in St. Louis, Missouri.[36] The first conference was called Biodevastation 7 and was organized by a coalition of U.S.-based activist groups and organizations opposed to genetic engineering.[37] The three-day event was held on a community college campus, and registration fees were set low (sixty dollars per person) to encourage attendance. Although the size of the crowd varied daily, about 150–180 people attended. Many of the speakers knew one another well enough to refer to one another by name.

The conference had the feeling of an informal teach-in, with most of the men dressed in jeans, sneakers or Birkenstocks, and T-shirts and the women clad in skirts and blouses. A group of twenty-five or so people arrived on bicycles as part of an organized bicycle caravan headed to Washington, D.C., to draw attention to the genetic engineering issue. Bicycles were everywhere, some wildly decorated. One had a woman's wig tied onto the seat as a seat cover. A security guard milled around the conference area, but he did not seem particularly threatening.

Core themes of the conference included environmental racism, "biopiracy," plant patenting and the commodification of life, and the corporate control of agriculture. Despite the casual feel of the meeting, the tone was serious, and many of the participants' comments reflected a deep suspicion and mistrust of corporations, U.S. government regulatory agencies, and even large research universities. One could detect a real sense of anger and outrage toward these institutions as people discussed issues, asked questions, and spoke passionately about the wrongs being done to society and the environment in the names of science, progress, and business.

On Saturday night, preceding the evening's three-hour plenary, African American drummers played African-style music, and eight colorfully dressed women performed African dances. After half an hour, the plenary began. The theme was race and environmental injustice. Five people spoke, including a Mexican farmworker–organizer from El Paso, Texas; an African American physician who ran the Connecticut Coalition for Environmental Justice; the director of the Institute for Science in Society in the United Kingdom; someone from the Federation of Southern Cooperatives; and Michael Hansen from Consumers' International. The speakers presented in a very informal manner, seemingly "off the cuff" and with no audiovisuals. The moderator, an African American woman in her midforties, opened the session by noting that she would follow a "feminist process," explaining to the audience, "If five white men have their hands up, and I skip over you to allow a woman and/or person of color to speak, that is because I am trying to get more voices heard." Clearly a skilled public speaker, the moderator sought to make the audience feel safe to ask questions and voice concerns. Indeed, many people participated in the lively discussion that followed.

The second conference, organized by the World Agricultural Forum (WAF), took place at the Hyatt Hotel in downtown St. Louis Union Station. Officially titled "A New Age in Agriculture: Working Together to Create the Future and Dismantle the Barriers," it attracted a very different group of participants and reflected a very different perspective on biotechnology and world agriculture. At a cost of six hundred to a thousand dollars a head, most attendees hailed from industry, farm organizations, governments, international organizations, and large foundations. Security at the conference was remarkably tight, and one needed an official, WAF-issued ID to enter. Virtually everyone at the conference wore an expensive-looking business suit and tie, and I felt conspicuously out of place in my forty-five-dollar Bergner's pantsuit. Dozens of policemen dressed in full riot gear stood outside the hotel, waiting for someone to disrupt the meeting.

The grand ballroom housing the plenary had a large stage framed by two enormous screens portraying a magnified image of the speakers. Directly

behind the podium stood another gigantic screen on which the organizers had projected an eye-catching collage. On the left side of the collage,
a large oil truck flanked a colorful market scene in Africa. In the middle,
an old black man sold vegetables. To his right, a group of Asian women
knelt and weeded fields in front of a tractor, a barge carrying big con-
tainers sailed on the open sea, and a modern skyscraper towered in the
background. The message could not have been clearer: tradition versus
modernity. Another message, reflected in the ships carrying the contain-
ers, suggested the significance of trade.

The four keynote speakers at the conference included Lennert Bage,
president of the International Fund for Agriculture and Development;
Norman Borlaug, a world famous plant breeder who had won a Nobel
Peace Prize for his work in the Green Revolution; David Raisbeck, vice-
chairman of Cargill, Inc.; and Pedro Sanchez, co-chair of the United
Nations Task Force on Hunger. Borlaug gave an impassioned speech about
the rapid growth of population in conjunction with lagging world food
supplies. In his view, the world could not wait for traditional methods
of plant breeding to solve the problem of malnutrition when the tools of
modern biotechnology were already in our hands. By the end of his talk,
Borlaug was speaking like a preacher and pleading with the audience to
"do something immediately!" to remove the obstacles to these technologies. With his whole body shaking, Borlaug lambasted critics for their role
in this world-food-crisis-in-the-making and proclaimed, "The precautionary approach is a disaster." He walked away to a thunder of applause.

The vice-chairman of Cargill and the other keynote speakers followed
with their own well-prepared, technologically sophisticated, and meticu-
lously delivered speeches. The last part of the morning involved two hour-
long roundtables organized like *Meet the Press*. Conference planners had
set up a large table in the middle of the room and positioned the audience
around them in concentric circles. A professional cameraman and light-
ing crew skillfully projected an image of the eight or so speakers onto the
screen as the camera gracefully moved around the panel. During these
roundtables, each moderator had the opportunity to pose questions and
guide the discussion; questions from the audience were neither invited nor
raised. The atmosphere felt professional, high-powered, and corporate.

The dominant discourse at the conference linked the end of world
hunger with removal of barriers to trade, as if it were self-evident that
reducing trade barriers would lead to greater food production and better access to food for the poor. Indeed, not until one of the few NGO
representatives present at the meeting raised the possibility that some

would lose in this process and that specific economic interests were push-ing for the trade opening associated with the Doha round of the WTO, did anyone challenge the view that increased global trade might not pro-vide benefits to everyone.[38] While some tensions over North–South trade emerged during the discussions, the atmosphere remained cordial, and there seemed to be a general consensus that more and better technology offered the solution to the world's food problems.

On the ride home from St. Louis, my head was spinning with these two experiences, and the different *Lebenswelten,* or lifeworlds, they rep-resented. I could not stop thinking about all of the assumptions and ideas that seemed to be taken for granted in these two worlds and how remark-ably different these two groups' frames of reference were. I had attended these conferences because William Munro and I were studying the politi-cal controversy over biotechnology, but I had not expected to walk out of them thinking so much about people's worldviews and culture. The image of these cognitively and culturally distinct worlds stayed with us for a long time and seemed to offer important insights into the nature and persistence of the agricultural biotechnology controversy. Several years later and half a world away, a South African activist reinforced our in-tuition. Asked during an interview why proponents and opponents had been unable to find common ground for debate, she paused and replied, "Just two very different ways of seeing the world."[39]

CHAPTER ONE

Precursors to Protest

In advanced capitalist economies, new technologies are developed and marketed on a daily basis. Generally, they do not spark widespread social resistance or protest, and we do not consider them to be infused with politics. The "biotechnological revolution" that introduced transgenic technologies in agriculture, however, unleashed a firestorm of opposition that took its proponents by surprise and refused to go away. Seed technologies lack a high public profile, and seed scientists tend to see themselves as contributing, in the quiet order of their laboratories, to age-old traditions of genetic manipulation. What conditions, then, catapulted their work into a global whirlpool of protest and conflict?

To understand why this issue exploded into a full-blown social controversy, it is important to recognize that the biotechnological revolution was not an esoteric and specialized process taking place on the fringes of the capitalist industrial economy. Rather, it lay at the heart of broad structural changes occurring in advanced capitalist societies and in the global economy during the second half of the twentieth century. In this chapter, we outline three developments key to the emergence and evolution of the controversy over agricultural biotechnology. One was the dramatic advances in science and technology associated with the shift to a "knowledge-based" economy; these advances, in turn, created the conditions for the development of the technology. Second, state approaches to regulating the economy reflected an ideological and political turn toward neoliberal globalization. Together, these developments laid the foundation for a highly dynamic, new global industry rooted in private-property-based technoscience.

The third development represented, to a large extent, a reaction to the first two developments—or more specifically to the societal and environmental changes they were believed to bring. It involved the emergence of a panoply of new social movements concerned with the risks and quality of life associated with late modernity on the one hand and with the negative manifestations of the "global development project" on the other. These movements represented the precursors to the anti-biotechnology movement: they focused on many closely related "predecessor" issues, were motivated by similar sensibilities, and established some of the local and transnational social networks that would later become the springboards for anti-GMO activism. New transgenic technologies were birthed into this world of growing social activism.

Biotechnology and the Knowledge Economy

In the decades following World War II, the economic core of advanced capitalist societies shifted away from industrial manufacturing as information-based industries, services, and high-technology enterprises rose to predominance. In the early 1970s, Daniel Bell introduced the concept of a "knowledge-based society" to characterize fundamental changes in the social and economic organization of industrialized societies, from industrialism to postindustrialism. This term captured the increasing power of knowledge within key economic sectors. At the heart of this shift lay dramatic technological developments driven by advances in the information sciences (information and communication technologies) and the biosciences (biotechnologies).[1]

One key trend in the development of knowledge-based economies was the increasing economic importance of the biological sciences, especially molecular biology and genetics.[2] For molecular biologists, the defining moment was James Watson and Francis Crick's 1953 discovery of the DNA double helix, which described the basic genetic structure of most living organisms. As knowledge of the genetic structure grew, the possibilities of "rewriting" or "editing" that structure captured the imagination of many cellular and molecular biologists. The potential implications of not only *knowing* the nature of life but also *intervening* in its production and design were clearly immense. For some, such intervention yielded the promise of creating "synthetic forms of life."[3] For others, it yielded the promise of engineering human beings to possess certain traits. In effect, the development of these new genetic technologies was infected with what Lily Kay calls a "molecular technological utopianism," which inspired a quickening sense of excitement among many scientists, entrepreneurs, and government officials.[4] Governments saw the potential of genetic intervention not only for stimulating economic growth but also for solving a variety of social problems across a number of public policy domains, including medicine, nutrition, and reproduction. Industry envisioned the ability to develop a wide range of new products and revenue streams. Through the 1960s and 1970s, the private sector invested growing amounts of money in the generation of this useful knowledge through funding for research, and governments elaborated new intellectual property rights (IPR) regimes that would protect investments in science and informatics. These developments, in turn, led to the explosion of complex scientific–industrial projects such as stem cell research and molecular biological research in the 1970s and 1980s, and the Human Genome Project, which sought to map human DNA in its entirety, in the 1990s. They were also the seedbed for modern agricultural biotechnology.

During this period, the combined pressures of the cold war, U.S. hegemony, and economic competitiveness in the emerging knowledge-based economy led advanced industrial states to adopt policy frameworks aimed at promoting science. This trend was especially pronounced in the United States, which had experienced a powerful upsurge of scientific development during World War II, much of it associated with the incipient arms race (especially nuclear technology) but also with the industrial production of penicillin. In 1950, most basic science research took place in university settings with strong public funding. Starting in the 1970s, however, private-sector research-and-development (R&D) expenditures rose dramatically and began to outstrip federal funding. Much of this money came from pharmaceutical companies drawing on advances in biochemistry to build their drug empires. Also, following the development of gene-splicing techniques in 1973, venture capital began to flow into the new biotechnology sector, fueling an explosion of small start-up companies led by scientist–entrepreneurs (see chapter 2). By the early 1980s, the private sector dominated R&D funding, and federal spending on basic science had dwindled considerably.[5]

The upsurge in biotechnology after World War II was thus associated with new dynamics at the heart of leading industrial economies, in which modernization, scientific development, and economic growth were understood as intrinsically related to one another through the promotion of technological change. In countries seeking to get ahead in this "second, science-driven industrial revolution," questions of how scientific knowledge could and should be nurtured, promoted, and used for the public good became an important public policy issue.[6] Though the ensuing debates were shaped by national political, cultural, and institutional contexts, they shared two basic orientations. One was a belief in the central role of technoscience in producing progress at both national and global levels. The other, as we outline below, was an important shift in thinking about how modern economies should work.

A New Approach to the Economy

The rising power of private interests over the scientific research domain did not take place at the expense of the state's interest in managing technology development; rather, that power was *facilitated* by shifts in government policy frameworks. Those policy frameworks were informed by governments' interests in promoting national economic growth and competitiveness in the postwar international economy. In the postwar world, the United States' hegemony gave the country a strong leadership position in shaping the competitive structures of an increasingly integrated global economy. Starting in the 1970s, the United States embraced a neoliberal economic doctrine at home and promoted it vigorously abroad. This doctrine stressed open markets, free trade, private property rights, and individual initiative. It promoted a stringent, market-oriented political order that emphasized a very different role for the state in managing economic activity, one in which public spending was tightly constrained and the state supported, rather than led, private initiatives. According to neoliberal theory, states should pursue "market-driven solutions" to the challenge of economic growth through policies emphasizing weak or permissive regulation for business activities, reduced or less accessible public services, lower tax burdens, and flexible labor markets.

Across Europe and North America, states undertook neoliberalizing strategies along these lines, although they did so unevenly. The particular mix of measures they adopted was shaped by national political and economic cultures, the balance of local social forces, and the perceived strategic imperatives of competing in an increasingly global economy. Since one of the keystone principles of neoliberal theory was a belief in free trade, neoliberalizing states needed to structure their national economies in ways that made them both internationally competitive and open to the flow of goods. However, open markets made them vulnerable to domination by stronger, more efficient, or more technologically advanced competitors. In an effort to overcome this vulnerability, states looked increasingly to international institutions to provide the rules for international economic interaction and to develop policy frameworks that were "harmonized" across borders. They also facilitated transnational flows of finance capital to fund their economic development projects.

The United States, drawing on its status as hegemon and eager to promote the global leadership of U.S.-based capital, took the lead in promoting this political–economic order. Consistent with its new ideological orientation, it began to advance a deregulatory sensibility in Washington, D.C. During the Carter administration, policymakers concerned about the economic slowdown associated with the oil crisis feared that zealous regulation of innovation might be inflationary.[7] Subsequently, the Reagan administration made economic "efficiency" and the compliance costs to corporations key considerations when evaluating new and existing regulatory policies. The government scaled back the capacity of its regulatory agencies considerably by cutting budgets, staffing, and research capacity, and it began to draw more heavily on external sources of knowledge that were increasingly provided by the private sector.[8] At the same time, the administration responded affirmatively to lobbying from research-intensive industries to subsidize private research efforts through a tax credit, making this incentive a permanent feature of the corporate tax code. In short, the U.S. government actively enabled private science rather than strengthening investments in public science.

Although other northern countries varied in their approach to neoliberal restructuring, they all adopted broadly similar policies to secure a competitive position in the rapidly globalizing and knowledge-driven economy. Over time, they came to articulate their strategies around the new concept of "innovation," seen as the driving force of economic growth and competitiveness. Innovation was defined not in terms of the generation of new knowledge, or in terms of the application of ideas in science and technology, but in terms of the application of knowledge in economic production in order to promote competitiveness in the modern capitalist economy. It was in pursuit of innovation-based international competitiveness that governments sought to establish supportive policy frameworks for privately funded, product-oriented science R&D. To do so, they subsidized research, established tax incentives for private innovation, created extensive property rights protections for innovators, and designed permissive regulatory regimes for technological innovation.

Furthermore, states pursued neoliberal policies even though that pursuit involved submitting themselves to international regimes that constrained their sovereignty. Thus, the state itself was restructured through the construction of international institutions for global economic governance, especially with respect to promoting free trade. Again under the leadership of the United States, agencies such as the World Bank, the International Monetary Fund (IMF), and regional development banks pushed international agreements on trade rules, foreign aid and lending practices, and development strategies that aimed to pry open national markets to the free flow of an increasingly wide range of products (from automobiles to foodstuffs to computer software). These efforts culminated in a regime of global economic governance that focused on the World Trade Organization.

This governance regime had three key components that helped shape policies relevant to agricultural biotechnology. One was a concerted push to liberalize the agricultural markets of developing countries by reducing import quotas, tariffs, and other barriers to trade. This process of liberalization was the central agenda of the Uruguay Round of the General Agreement on Tariffs and Trade (GATT), which began in 1986 and ushered in the World Trade Organization in 1995. Many critics saw GATT as a tool of U.S. hegemony and an instrument to develop new markets for U.S. products, including food.[9] Nevertheless, GATT and the WTO came to provide a locus for intense agricultural trade wars between the United States and the European Union, including U.S. challenges in the late 1990s to European prohibitions on the importation of hormone-treated beef and genetically modified organisms. They also provided a means by which countries from the global South could challenge sustained agricultural subsidies in the United States, Europe, and Japan.

The second key component of this global economic regime was the effort to harmonize national regulatory environments by locating rule-making authority in multilateral organizations. This component was especially relevant to governments and corporations eager to promote a global agrofood system through increased trade in agricultural products. Globalized food supply chains cannot work smoothly and efficiently if countries regulate food safety on the basis of different types of risk or use different local standards. Thus, in 1962, the Food and Agriculture Organization and the World Health Organization established the Codex Alimentarius Commission to harmonize food standards in ways that both protected the health of consumers and ensured fair practices in the food trade. Although the codex provided voluntary guidelines for members, it gained greater power to discipline states after the establishment of the WTO, which had the power to make binding rules on member states and to mandate that states employ regulations that are the least "trade restrictive." Under the new food safety regime, the basic requirement of regulatory policy was that it should be based on "good science," not on "irrelevant" criteria such as aesthetics, culture, or potential environm tal damage, which might inhibit the free flow of goods.[10] Only in the absence of existing scientific evidence could a country take an independent precautionary approach to trade involving agricultural products.

The third component of the global governance regime involved an effort to strengthen the protection of private property rights in an increasingly integrated global economy. As already noted, the central idea driving this effort was the conviction that economic growth is best promoted through privately funded competitive innovation and that the appropriate role of the state is to provide a facilitating environment for such innovation. This could be done by establishing strong legal and institutional mechanisms (e.g., patents, trademarks, and copyrights) for protecting intellectual property rights in such "creations of the mind" as inventions, designs, new plant varieties, and works of art and literature. Governments and businesses regarded IPR as essential because the knowledge-based economy is research intensive, and without strong protection for their hefty investments in R&D, private entrepreneurs would not have the economic security to invest in often risky innovations. Moreover, in an increasingly integrated and trade-based global economy, such protections had to be internationally harmonized and enforced as much as possible. This led to the establishment in 1994 of the international Agreement on Trade-Related Aspects of Intellectual Property Rights (TRIPS), administered by the WTO, which aimed to provide enforceable global standards for protecting IPRs in a wide range of high-technology fields. As Article 27.1 of the TRIPS agreement states, "Patents shall be available for any inventions, *whether products or processes,* in all fields of technology, provided that they are new, involve an inventive step and are capable of industrial application" (emphasis added).[11]

Efforts to build this regime reflected a confluence of interests among developed industrial countries, international financial institutions, and multinational corporations—converging into what Philip McMichael has called the Globalization Project.[12] In the agricultural sector, this implied a modern agrofood system embedded in global rather than local trade networks, markets, and commodity chains. States came under pressure to deregulate agriculture and food markets, though they did so quite unevenly: while international donors and aid agencies forced poor countries to open their markets substantially, rich countries tended to maintain protections for their farmers, often through large export subsidies. As a result, international trade in agricultural products increased dramatically. Trade liberalization subjected farmers to new market realities, including

petitors, consumer tastes, and quality standards. At the
gn direct investment in agriculture expanded, especially
)nal corporations established foreign affiliates around the
:ct of this process was a global concentration of the seed
large multinational companies. To consolidate their competitive position, these companies increased their levels of R&D in seed
technology and vigorously called for the protection of "intellectual property rights for bioengineered seeds, plants, and animals as the fundamental basis for innovation."[13] Thus, they laid the institutional foundation
for the "GM revolution" in a neoliberal era.

New Social Movements

As we suggested in the Introduction, the development of biotechnology in
the 1970s and 1980s involved extensive efforts to expand human beings'
control over nature, including the production of novel life forms. Thus,
the profoundly transformative implications of such far-reaching efforts
would inevitably reverberate through society. In the first place, these efforts promised to alter not only nature but also social institutions, cultures,
and norms of accountability. For instance, developments in reproductive
technology, recombinant DNA, and stem cell research raised fundamental
questions about the ontological status of the human self and its place in
larger natural, social, and political orders. As a result, these developments
challenged fundamental values and ethical principles in ways that made
the formulation of policy politically sensitive.[14] More prosaically, the
brave new worlds of scientific and technological innovation raised new
questions about health, safety, and acceptable risk to consumers and the
environment. In a highly influential book, the German sociologist Ulrich
Beck argued that the powerful technologies of the post–World War II
era—nuclear and chemical as well as biotechnological—had unleashed
new threats to human and ecological well-being that were not only unprecedented in the history of industrial development but also qualitatively
more threatening because they were unbounded in time and space.[15]
Nuclear fallout, for instance, could not be spatially contained. Similar unbounded risks were associated with chemical technologies, which were
responsible for acid rain, the hole in the ozone layer, toxic waste, and
perhaps global warming. The cumulative effect, according to Beck, was
that citizens today live in a "risk society" in which the potential threats
of technological advances to human and ecological well-being cannot be
fully known and, once unleashed, could not be fully managed.

Even the prospect of such threats was deeply unsettling to some citizens and led them to organize collectively to address these risks. Beginning in the 1960s and 1970s, a plethora of "new social movements" emerged in Europe and the United States. Some of these movements were the product of post–World War II affluence and drew upon and elaborated a complex and variegated criticism of late capitalist society. They organized around human rights and citizens' rights issues, such as peace, nuclear power, the environment, women's rights, and sustainable agriculture. Although these movements mobilized significantly different ideologies, they marked an important, and in some ways novel, engagement with the institutional and organizational features of social power associated with advanced capitalism and the new knowledge economy.

One of the most overarching and profound concerns these movements voiced was the social and environmental costs of postindustrial society. A critical issue was the increasingly obvious capacity of humans to destroy the earth and their own well-being with new technologies: the threat of obliteration posed by nuclear power and chemical warfare, the increasing scarcity of nonrenewable natural resources such as oil and coal, the large industrial interventions in local and global environments, and the massive impacts of chemical-dependent industrial agriculture on ecosystems. Though these problems were large, systemic, and often environmental, they had a very tangible presence in citizens' own lived experiences of increased pollution (especially toxins and acid rain), the prospect of nuclear meltdowns, the crumbling of urban environments and infrastructures, and the ever-present threat of cancer. In addition, the tendency of governments and regulatory agencies to displace the social and environmental costs of this economic development model to poor communities, both domestically and globally, motivated activists to work on issues related to human rights, environmental justice, appropriate technologies, and sustainable livelihoods and development.

The anti-nuclear movement distilled the sensibilities of these movements in a particularly clear way by focusing concerns about peace, nuclear power, toxic waste, accountability, and quality of life directly on a potentially apocalyptic and unnecessary technology. The growth of the anti-nuclear movement reflected a growing perception that the organization of modern industrial societies was imposing critical limits on the quality of life of ordinary people; what was the point of being materially secure (under the welfare state) if you lived under the constant threat of being destroyed by nuclear fallout, toxic waste, and other forms of pollution? As a result, these activists became energized by a growing appreciation that, for all the liberating effects of a modern knowledge-based society, life in the modern world was hedged with very potent and quite terrifying risks and dangers. What is more, these threats were themselves an outcome of the institutional organization of an economic growth model based on both technology and private property.

Given that the risks associated with late modernity were seen to be potentially threatening to the human species, concerned citizens began to ask whether the existing model of social organization was really the best institutional arrangement for securing the quality of life (and selfhood) of citizens, as well as the future of society. Not least, they began to place a high premium on how society would decide what levels and kinds of risks are acceptable. Industry and government promotion of technologies such as agrochemicals, industrial agriculture, and nuclear power raised critical questions about accountability, responsibility, and democratic voice/choice in the knowledge economy. A pressing question was whether the dominant technology- and growth-based model of social organization was capable of producing a "good society," one that wouldn't destroy itself in its quest for progress and profits. In the words of Andy Kimbrell, a long-time anti-genetic-engineering activist,

> There was a change, a significant change. . . . What you had were people who began to say, "Well, wait a minute; we want to question the whole industrial paradigm." . . . Part and parcel of what we tried to do in the '60s, whether it be war or something else, [was] not [to] allow other people to create these huge systems where we weren't actually understanding our responsibility.[16]

Somewhat different, though related, questions about responsibility and democratic accountability were also raised about North–South relationships in the postwar global economy. By the early 1970s, a number of people and organizations had become highly critical of the dominant development paradigm. These grassroots-oriented development organizations had reached the conclusion that the "development" policies and projects being promoted by northern states, international aid agencies, and development banks on behalf of the "third world" were not improving conditions for the poor but actually worsening them. Global development institutions, such as the World Bank and the International Monetary Fund, and bilateral aid agencies, such as the U.S. Agency for International Development, tended to finance large-scale infrastructural projects such as dams, highways, and electricity systems that sustained and benefited third world elites and exacerbated social and economic inequalities in those countries. What is more, these projects tended to be loan- rather than aid-based, meaning that they inevitably generated large debts on the part of recipient countries. These same grassroots individuals also identified negative consequences of North–South "technology transfers," including the hybrid wheat and rice varieties that formed the backbone of the Green Revolution. They saw these technologies as offering simple technological fixes that were too large-scale, too expensive, and fundamentally inappropriate for the people they were ostensibly designed to help.

In the 1980s, organizations concerned about the impact of the dominant development paradigm on health, biodiversity, and other aspects of the environment grew dramatically in size and number and became increasingly networked with one another. These organizations mobilized around such issues as the environmental disasters of Bhopal and Chernobyl, the whaling industry, the sale of baby formula in developing countries, and large-scale dam projects. Some of these organizations were international, such as Greenpeace, Friends of the Earth, the International Organization of Consumers Unions, and the Pesticide Action Network; others were national. In eastern Europe, a wave of nongovernmental organizations and local social movements emerged in the context of the crisis of state socialism, de-industrialization, and urban pollution. Similar processes took place in Latin America, where civil society was strengthened in the 1980s as a result of the area's democratic transitions, and in Africa, where the capacity of many states declined in association with both rampant corruption and neoliberal restructuring. In South and Southeast Asia, growing movements drew on long traditions of community organization.

Though these organizations had their own specific concerns and campaigns, they shared a focus on the challenges posed to "sustainable development" by the technology-driven development paradigm of the Globalization Project. The increasing access of civil society organizations to multilateral deliberations over sustainable development at the United Nations facilitated networking and coalition building among these activist groups. A key moment was the 1992 United Nations Conference on Environment and Development (frequently called the Rio Summit), which provided the launching pad for the Convention on Biological Diversity (CBD) and its Cartagena Protocol on Biosafety, which aimed to establish a regulatory regime for the cross-border flow of transgenic organisms (discussed in chapter 6).[17] For many activists, especially in the global South, the modern agroindustrial system was inherently unsustainable, in part because they believed it posed profound and potentially irreversible threats to biodiversity.

In the wake of the Rio Summit a number of new, critically minded organizations emerged across developing countries, organizing around issues of sustainable development, farmers' and community rights, the protection of biodiversity, and environmental justice. Over time, these critical civil society voices became an increasingly noticeable and regular presence in the negotiation of global environmental governance institutions, especially within the United Nations system. At the same time, those institutions themselves became a vehicle for promoting the emergence and consolidation of new similar organizations, thus helping to catalyze opposition to GMOs.

Civil society activism and networking played a similarly important role in the international negotiations to restructure global economic policies, especially the framework for a global system of agricultural trade and the imposition of economic adjustment programs and aid conditionalities on developing countries. Though civil society organizations and activist groups did not participate directly in these multilateral negotiations, many in both the North and the South were disquieted by the effects they believed that neoliberal deregulation would have on the environments and livelihoods of those who were most vulnerable around the globe. Many were also angered at the austerity programs and debt burdens imposed on poor countries by the World Bank and the IMF, which they saw as hurting the world's poor majorities. In the global South, the rising prevalence of popular protest and "IMF riots" fueled solidarities both between northern and southern NGOs and among organizations based in the global South. Through years of transnational networking and planning, civil society groups mobilized several broad anti-globalization campaigns in northern countries, such as the "Fifty Years Is Enough" campaign to close down the World Bank and the "Jubilee 2000" campaign for universal debt relief. This rising tide of mobilization came of age in December 1999 with the extraordinary protest against the WTO Ministerial Conference in Seattle.

Although agricultural biotechnology was not a central issue motivating the Seattle action, it had a visible presence in the form of street theater and costumes and played a bridging role between concerns about the globalization paradigm's threats to biodiversity and the domination of global agriculture by a small group of northern companies.[18] For some of the activists in Seattle, agricultural biotechnology represented one aspect of a globalization process that was imperialistic, domineering, and fundamentally inequitable. They also saw it as a focus for effective push-back by civil society. In sum, the battle in Seattle highlighted the rising role of civil society networks in contesting the neoliberal international order in which the biotechnology industry was nestled.

From Private Science to Biotech Battles

As Scott Prudham has noted, "Biotechnology and genetic engineering do not simply happen but are instead called forth by particular cultural and institutional processes."[19] As we have shown in this chapter, at the broad macrohistorical level, these processes involved the consolidation of an integrated and global economic system that was increasingly based on the growth in knowledge-intensive industries and was driven by a logic of competitive innovation in science and technology. Heavily dependent on the establishment of intellectual property rights and their extension into new economic and geographical realms, this system was consolidated under the linked dynamics of U.S. hegemony and neoliberalism. It gave private interests a rising role along with increased power in the domain of scientific research, and it established a supportive environment for private innovation. It also expounded a profoundly optimistic vision of modern societies in which technoscience would provide the foundation for economic growth, the basis of national competitiveness, and the solutions to a wide range of social problems.

Yet, at the same time that the new biotechnologies emerged within a particular historical context, the processes that brought these technologies into existence also called forth a new set of social actors, criticisms, and coalitions to contest them. The risks, dangers, and ills associated with late capitalist development and the more contingent processes of neoliberal globalization created environmental problems, economic policies, and global power relations that were ripe for protest and produced a cadre of people who were motivated to take action against them. The rise of information- and knowledge-based industries created citizens' demand for increasing amounts of information about technologies and policies that could affect their lives; however, this was information that companies were disinclined to share, since it affected their profits and competitiveness. Moreover, because the vast preponderance of technoscientific capacity lay in the global North, citizens in poor countries feared that external technologies would be imposed upon them. For many citizens, then, the model of modernization based on technoscientific innovation in a neoliberal age was potentially destructive and disempowering rather than progressive and liberating. The critical readings these social actors gave to these phenomena created a fertile field in which opposition to genetically modified organisms would grow, and they helped to generate a dialectical tension between the proponents and opponents of biotechnology.

CHAPTER TWO

Creating an Industry Actor

I believe—and many hard-headed scientists agree with me—that
with the new biotechnology, almost anything . . . can be achieved:
new organisms, new limbs and kidneys, new treatments for disease,
new ways of controlling pests, crops which produce their own herbicides. . . . Whole new industries will sell products that today cannot
even be imagined, let alone made.

—John W. Hanley, chairman, Monsanto Company, 1982

It is appropriate for us to take a long look at our own industry
[chemicals]. . . . And when we do, we see an industry that has made
major contributions to the quality of life in the United States and
in the world. We see an industry that is growing and prospering. We
see an industry that is a keystone to the nation's economy and its
well-being. . . . What we don't see . . . is the public esteem that used
to accompany our achievements. In its place, . . . we find criticism,
suspicion, and hostility directed toward chemicals and the chemical
industry.

—Dr. Louis Fernandez, chairman, Monsanto Company, August 1984

In the 1970s, two phenomena occurring in two very different sectors
of society helped to forge a new industry with enormous consequences
for the future of agriculture. One of these phenomena occurred in the
sphere of science. In 1973, building on James Watson and Francis Crick's
pathbreaking work revealing the double helix structure of DNA, Herbert
Boyer, a biochemist at the University of California at San Francisco, and
Stanley Cohen, a professor of medicine at Stanford University, developed

a way to splice genes from one organism into another, effectively cross-ing the "species barrier." With this remarkable discovery of recombinant DNA (rDNA), Boyer and Cohen, along with others who had been study-ing genes and gene functions for years, opened up a brave new world in the biosciences. For those who were a part of this scientific community, nothing could have been more exhilarating.

The other phenomenon—or more accurately, confluence of phenomena—occurred in the world of business. For the first time in history, a new environmental consciousness swept through many countries and helped to create a growing distrust and criticism of industries that were known to be major polluters of the environment. Prime among them were the chemical and petrochemical industries. As indicated by Louis Fernandez in the second epigraph above, no longer were the makers of chemicals and chemical products received by the public with enthusiasm; instead, they were greeted with hostility and criticism. In this increasingly nega-tive political climate, the large multinational firms that made up this and related industries (e.g., agrochemicals, petrochemicals) realized it was time for reinvention.

This chapter tells the story of how these once-separate communities, cultures, and worlds of the biological sciences, on the one hand, and of multinational chemical, energy, and pharmaceutical corporations, on the other, came together to form a particular kind of cultural–economic actor. This actor comprised a dozen or so multibillion dollar corporations that shared a rationality, set of norms, and commonsense ideas about what they needed to do to be successful, along with a certain way of looking at the world. At the core of this lifeworld was a belief in the fundamentally positive nature of science and technology and an adherence to the idea that a scientific perspective, which relied on "hard facts" and empirical evidence rather than on religion, value judgments, or emotion, was quin-tessentially rational. This belief was intimately attached to a value system and mindset in which success and status were defined in terms of firm growth, profits (and, after 1980, shareholder value), and achievement of a competitive edge.

On the basis of this culture and worldview, this group of corporations set out over the course of the 1980s and 1990s to construct an "inte-grated life science" industry. They did this by hiring large numbers of scientists and acquiring many of the small entrepreneurial start-ups that had initiated the biotechnology industry in the late 1970s. Convinced of the emerging biotechnologies' potential synergies among agricultural, pharmaceutical, nutritional, and medical applications and attracted to the possibility of reinventing themselves as cleaner, greener, and health-oriented companies, these multinationals brought both their vast economic resources and everyday ways of acting, thinking, and operating in the world to bear on the task of industrial transformation. Over time, they created a virtual juggernaut of biotechnology.

Nevertheless, this juggernaut was characterized by crucial blind spots, vulnerabilities, and culturally shaped behaviors that ultimately created openings for the opposition of anti-biotechnology activists. For example, many of the high-level executives of these multinational corporations believed that whatever opposition arose to these new technologies could be effectively managed through a (corporate-designed) policy of government regulation and public education. From their perspective, the naysayers who failed to acknowledge the benefits of biotechnology were simply anti-technology "neo-Luddites" or environmental extremists who were unlikely to garner much sympathy among policymakers or the general public. For their part, the scientists in these companies (and some of the managers) were so convinced of the transformative power and social and environmental benefits of genetic engineering that they could not imagine that social opposition to these technologies would seriously grow or become significant in some other way. Consequently, both groups tended to discount the legitimacy and the import of the opposition.

A further vulnerability lay in the large investments and long lead times necessary to develop products and move them through the patenting and regulatory processes, which placed enormous pressure on firms to get their products into the marketplace quickly. Because a significant delay could be devastating, U.S. firms aggressively sought to secure a favorable business environment. However, the way they went about this was constrained by their culturally informed understandings of the scope and nature of what sociologists call their "organizational field."[1] By defining their central stakeholders as (mainly U.S.) farmers who bought their products and the government food safety authorities who regulated them, these firms failed to adequately gauge the importance of cooperation at the downstream end of the commodity chain, comprising food consumers. These large firms' business cultures and competitive strategies also contributed to certain behaviors that antagonized the opposition. For example, with these firms living in a world in which Wall Street was increasingly their chief judge and referent, a stock feature of their business plans was the need to establish intellectual property rights over commercially valuable research.

Although many multinational corporations bought into the idea (and *ideal*) of a "life science" model in the 1980s and 1990s, they did not all go about trying to create such an enterprise in the same way. Their variation in large part derived from differences in their business strategies. Firm strategies were shaped by the internal cultures of these companies as well as by the personal proclivities of their top managers, especially their chief executive officers.[2] They were also affected by a firm's specific historical trajectory and its managers' perceptions of the economic prospects for different activities. In short, firms varied significantly in deciding how serious a commitment to make to the biosciences and how fast and hard to move on this commitment.

In analyzing how firm cultures affected the behavior of the companies investing in agricultural biotechnology, we pay particularly close attention in this chapter to one company, namely, Monsanto. Monsanto was one of the first large chemical companies to move into the biotechnology sector in the 1970s and rose to prominence as the industry's leader in the 1980s, a position it retains today. While many other firms were active developers of the technology and important actors in the industry, no other company invested as much time, money, and human resources in establishing its position in the industry. Nor did any other firm exert the same degree of influence over the fate of the industry and technology. Indeed, if there was one company whose name became virtually synonymous around the world with the term *GMO*, that firm was unquestionably Monsanto.

Industry Beginnings

The biotechnology industry actually began not with large multinational corporations but with a group of small, specialized firms called "new biotechnology firms" or "biotech start-ups."[3] These firms first came onto the scene in the mid-1970s and grew rapidly in number over the next half dozen or so years, reaching well over one hundred by 1982. Although few biotech start-ups managed to survive very long as independent companies, they played an important role in moving molecular biology, and molecular biologists, out of the university and into the sphere of industry. They also fueled tremendous excitement around the technology and generated many of the scientific advances that propelled the industry's takeoff.

The story of Genentech illustrates the process by which many of these new biotech firms got started. Shortly after Herbert Boyer and Stanley Cohen invented their technique for gene splicing, a young venture capitalist by the name of Robert Swanson read about the breakthrough in *Science* and became enthused by the commercial possibilities rDNA had to offer. Swanson approached Boyer with the idea of starting a business, and Boyer agreed to lend his lab, talent, and reputation to the endeavor. Within the year, Genentech was born.

Following on Genentech's heels were companies such as Genex, Biogen, Hybritech, Molecular Genetics, Calgene, Genetic Systems, and more. Virtually all combined the same three ingredients: a few entrepreneurially minded scientists who either wanted to leave the university or could be tempted to do so to ply their trade in industry, an MBA or two capable of turning an esoteric scientific development into an appealing business idea, and a group of investors who were interested in backing a high-tech, high-risk venture in exchange for the promise of large economic returns. With these ingredients in place, the biotechnology industry took root.[4]

By most accounts, the new biotech companies, and the people within them, were a highly dynamic and entrepreneurial group.[5] Several factors contributed to the sense of intensity and drive that characterized the cultures of many small biotech start-ups. One was the sheer excitement people felt as they engaged in the new world of molecular biology and the related biosciences and sought to apply them in the fields of medicine, pharmacology, and agriculture. Here was an arena in which the possibilities seemed endless, both in terms of the societal need for these products and in terms of their market demand. One of the first products that Genentech began working on, for example, was a genetically engineered bacterium that could produce insulin, for which annual market demand in the United States alone was estimated to be on the order of several billion dollars. Another California-based firm, Hybritech, focused its efforts on developing monoclonal antibodies that could be used to diagnose and treat disease; this, too, represented a fantastically large market.[6] With market potentials like these, it was hard for people in the industry *not* to feel excited.

The nature of competition in the new industry contributed to the adrenaline-filled atmosphere found in many biotech start-ups. In the initial phase of the industry's development, competition in the scientific and business world of genetic engineering revolved around identifying genes and their functions before anyone else did. Time was of the essence in the race for genetic discoveries, and scientists worked feverishly to reach the finish line before their colleagues in other research teams.[7] The reason scientific discovery took on such importance was straightforward: under U.S. patent law, being the "first to invent" something was a key litmus test for being awarded a patent. Once a company established intellectual property rights over a gene or genetic transformation, it could claim exclusionary rights to use it or license it to others, thereby ensuring a stream of revenue from the investment. Patents enabled firms to realize value in their biotechnology investments and at the same time to concretize that value in the property right itself. For smaller companies that were chronically short of research funds, the concretized value of an intellectual property right could mean survival as a business.

Yet no matter how much brainpower and effort these scientists and their business counterparts poured into their jobs and how successfully they staked their claims at the patent office, small biotech companies faced an uphill battle in keeping their businesses alive. Conducting research using the new molecular techniques was intrinsically expensive because of the high quotient of unknowns characterizing the science, the cost of the highly skilled personnel needed to do it, and the long-term nature of the research. Furthermore, most of the research taking place in the early phase of the industry's development was still primarily directed at testing out ideas rather than producing products. As a result, most companies did not have anything to sell in the marketplace for some time. While it was not difficult for a new company with a couple of distinguished scientists to interest some risk-oriented investors to support their endeavors for a couple of years, it *was* difficult to sustain that revenue stream. As the cofounder of one new biotech firm lamented, "It's hard to keep going back to Wall Street and say, 'We need more money. We need more money.' They just weren't going to do it anymore." What typically happened to firms like his, if they were lucky, was that a large corporation would say "Well, you've really invented something, and we have money; we'll help you finish." They made people an offer they couldn't refuse.[8] For many start-up owners, being bought out by a bigger company or having one purchase a large equity share in the small firm was their best hope for staying in business. Indeed, for some small start-up founders, and even more for the venture capitalists that funded them, being bought out by a large company was an explicit goal.

A Magnet for the Multinationals

Starting in the late 1970s, a group of large multinational corporations also took a growing interest in the emerging science of genetic engineering. Among them were major producers of chemicals and agrochemicals, such as Monsanto, Dow Chemical, and American Cyanamid; firms whose primary business was in pharmaceuticals, such as Eli Lilly, Sandoz, Ciba-Geigy, BASF; and firms that were primarily known as energy companies, such as Royal Dutch Shell and Occidental Petroleum. Because of the nature of their businesses and their ongoing investments in scientific research and development, these multinational firms had historically kept their eyes on the developments taking place in science and technology. Thus, when news of rDNA started spreading and dozens of small biotech start-ups began appearing, these companies were quick to take note.

To those at the helm of these large corporations, investing in new life sciences was seen as a way to put their companies on the cutting edge. John W. Hanley, the CEO of Monsanto at the time, explained,

> What made us want to . . . make major investments in genetic engineering, was [the belief] that everyone in seed technology and agricultural technology and in pharmaceuticals was really going to [have to] start all over again as a result of genetic engineering. The parameters of what genetic engineering could mean, even in those days, were so broad, so wide, and so deep that it just seemed a foregone conclusion that if one mastered genetic engineering, one would be at the forefront of the development of new products.[9]

As another industry participant recalled the moment: "Our company [Allied Chemical] saw this as many companies saw it, that biotechnology will be the business of the future. . . . All these companies—Ciba, Sandoz, Zeneca—they were all pretty traditional companies when it came to agriculture and chemicals. . . . Everybody saw biotechnology as a way to become nontraditional, and there was clearly some money to be made."[10] Many large companies were also powerfully attracted to the idea being spread by both scientists and the business press that key synergies were to be gained by bringing the sciences of chemistry, biology, molecular genetics, and pharmacology together under one roof. According to the proponents of this integrated life science model, basic research in the new life sciences could benefit a wide range of activities, from drug research and production to energy production and the transformation of animal and plant life. The heads of these corporations had only to seize these opportunities.

The chemical industry was particularly motivated to invest because of pressures emanating from its own central business, whose growth prospects on the eve of biotech's birth appeared quite dim. In the early 1960s, Rachel Carson's widely read anti-chemical treatise, *Silent Spring,* swept through the United States and helped to fuel a new environmental movement both at home and abroad. The effect of a burgeoning environmental consciousness on the political climate for chemical companies was palpable, as revealed in the second epigraph at the beginning of this chapter. Although industry executives would never admit it openly, public attitudes had so soured toward the chemical industry that some executives believed their businesses' future rested on moving away from "dirty" chemicals and into activities that the public would consider cleaner and greener. Given its widely touted potential to help ameliorate environmental problems, biotechnology offered an apparent way for chemical companies to reinvent themselves in a new, more environmentally sensitive guise.

Two additional concerns helped to push the chemical industry toward a biotechnology future. The first comprised a new group of economic pressures that beset the industry. During the 1950s and 1960s, large chemical companies had amassed substantial profits based on a wave of innovation in chemicals, plastics, and other kinds of synthetics. But by the 1970s, many of these activities no longer generated the same level of growth or profits. Patents were running out, and industry costs were on the rise as a result of new environmental regulations, a jump in energy costs, and a series of lawsuits brought against the industry. As a result, "profitability really went to hell in a handbasket," reported Al MacLachlan, a senior vice-president of DuPont at the time.[11] The second concern was psychological in nature and had to do with corporate executives' judgments about the direction their industries were taking. With each passing day, biotechnology looked more and more like the wave of the future. For firms that *failed* to invest and missed what turned out to be a crucial wave, the consequences could be catastrophic. Thus, some companies invested in the new biosciences as a defensive strategy in order to reduce the chance of losing their industry standing.

Gaining a Foothold

As suggested by their Fortune 500 names—DuPont, Abbott, Dow Chemical, BASF—the multinational corporations that moved into biotechnology in the 1980s were enormous business conglomerates, replete with massive R&D budgets, a wide range of economic interests, and worldwide scopes of operation. In 1982, DuPont, for example, was a thirty-three-billion-dollar company, with approximately ninety major businesses operating in over fifty countries.[12] At $2.6 billion in sales that same year, Abbott Laboratories was a major multinational pharmaceutical firm.[13] These were companies with millions of dollars to commit to a new economic arena. Within a few years, all of the multinational companies that were serious about this new endeavor had spent hundreds of millions of dollars.

Multinational corporations pursued four complementary strategies to gain a foothold in the area of biotechnology.[14] The first involved building up their own "in-house" research capacities, usually by hiring a highly distinguished scientist and giving him or her the task of pulling together a world-class research team to work for the company. When Monsanto decided to invest in biotechnology, for example, it spent $185 million on building a new state-of-the-art molecular biology lab in St. Louis, Missouri, and recruited Dr. Howard Schneiderman, then dean of the School of Biological Sciences at University of California Irvine, to come and head it.[15] As the company's new senior vice-president for research and development, Schneiderman was charged with the task of making Monsanto into "a significant world factor in molecular biology" and was given the funds to hire all the research scientists he needed to do it.[16] Following a similar strategy in 1982, DuPont hired Mark Pearson, then director of the Molecular Biology Laboratory at the National Cancer Institute, to run its new molecular biology research program. With the help of several sub-lieutenants, Pearson recruited several hundred new scientists to work on a wide range of pharmaceutical and agriculture-related projects.[17]

A second strategy these large firms employed was to establish long-term affiliations with university faculty and their laboratories. Such affiliations allowed a multinational corporation to keep a close eye on the cutting-edge biological research being conducted at universities while also giving them certain rights to license the patented discoveries that came out of these projects. Third, the multinational corporations (MNCs) contracted with small biotech firms to carry out specific research projects.[18] For a company like Monsanto or Allied, the advantages of contracting were that the MNC could find the expertise and production capacity it needed within an existing firm and not have to pay the cost of hiring the scientists who specialized in an area.[19] Finally, many large companies purchased equity shares in new biotechnology start-ups or bought them outright if they were particularly attractive. Buying an equity share in a start-up provided an MNC with access to a start-up's specialized areas of research and any economic windfalls it might enjoy if its scientists discovered something important or the company did well in the stock market or both. If a start-up was especially "hot," owned patents that the large firm coveted, or offered significant competition to an MNC, these large firms simply bought up the smaller company, acquiring its assets (including its patents) and eliminating the competition at the same time.

Meeting at the Altar: The Marriage of Bioscience and Business

In the process of creating (or re-creating) themselves as biotech companies, the biotech start-ups and large multinationals that invested in genetic engineering brought together two distinct worlds, worldviews, and cultures—those of the molecular biologists, plant scientists, and geneticists who came to work in the industry, and those of the people who ran or managed these large and small corporations. When they came together, each brought its own ways of thinking and acting and its own belief system to the union, and each influenced the other in important, if sometimes subtle, ways.

The World of the Scientists

As we have just observed, when large multinational corporations decided to move into biotechnology research, they often added significantly to their scientific ranks. Most of the scientists they hired came directly out of universities, where the bulk of molecular biology research was still being done. Small biotechnology companies also contributed to the flow of former university scientists into industry. Although many research scientists were initially reluctant to leave the world of the university with its strong reputation of neutrality and its clear standards of scientific excellence, they gradually overcame their reticence and began to carve out a new career path. Indeed, working in a company, especially a large one, offered professional scientists some significant advantages over a university post: far better financial compensation, the chance to pursue one's research interests without having to spend time writing grant proposals, and the opportunity to work with other first-rate colleagues on state-of-the-art research.[20]

For scientists who worked in the biotech industry, particularly in large and deep-pocketed firms like DuPont and Monsanto, the 1980s were a dream come true. There was money, there were fantastic colleagues, and there was the opportunity to pursue basic scientific research and publish it in the world's most highly respected scientific journals. What more could a scientist want? Dr. Ralph Quatrano was one of the scientists who took a job at DuPont in the mid-1980s.[21] After spending almost two decades as a botany professor at Oregon State University, Quatrano agreed to take a leave from OSU to help head up DuPont's plant research group. As Quatrano recalled, the buildup of scientific expertise at the Delaware-based company during his three-year stint there was swift and impressive. Starting from nothing, DuPont hired dozens of new researchers in his area alone and let them follow their noses. "It was like heaven, no budgets, whatever . . . you wanted to do," he reported, with more than a hint of nostalgia in his voice.

> I had probably the greatest group of fifteen PI's [principal investigators] I could ever possibly imagine. They were doing basic research. I did some of my best work in those three years with one or two postdocs.

There was a *Science* paper and a *Nature* paper. . . . I never had that
happen in academia![22]

To top it off, there was no pressure for product development, at least
during the first few years.

The degree of excitement that scientists felt as they forged the new
world of genetic engineering is hard to overestimate. In the years following the discovery of rDNA, scientists shared a powerful sense (one
that was picked up by the media) that this was a technology with virtually endless possibilities. According to the discourse of the day, genetic
engineering could permit fish to survive in freezing cold waters, produce
vine-ripened tomatoes that would not bruise on their way to market, and
yield more nutritional crops. The words of Mary Dell Chilton, one of the
pioneers of this new technology, reflected this widely shared optimism.
"The solutions are coming very fast now," proclaimed Chilton not long
after biotech's birth. "In three years, we'll be able to do anything [with
gene manipulations] that our imaginations will get us to."[23]

Chilton's assessment turned out to be a bit inflated. However, scientific
advances were not long in coming. In less than a decade, scientists successfully figured out how to produce human growth hormone synthetically, engineer a naturally occurring insect toxin called *Bacillus thuringiensis (Bt)* into plants, and make plants resistant to specific herbicides. The
thrill of being a part of such discoveries was intoxicating and only whet-
ted scientists' appetites to learn more. A long-time Monsanto scientist
described the feeling: "People like Rob, myself, Phil, who was the chief
scientist of the company . . . we were all kind of, you know, 'Let's do the
next big thing. What's the next front here? Where's the next mountain
to climb?'"[24]

Not many questions arose about the safety of the technology. In part
because they had been trained in the techniques and in part because they
felt confident in their own knowledge of the science, the vast majority
of industry scientists believed these techniques were safe for both humans and the environment. This belief only grew stronger as time went
on and those working with rDNA techniques became even more certain
about the "soundness" of the science. "This is the same stuff we've been
doing for years; it's just better, more efficient. It's inherently safer, at least
from all that we know of the science after the last fifteen years," noted
an expert in plant viruses.[25] Most scientists readily admitted that some
risks were involved, but they gauged those to be small. Furthermore, all
technologies carried risks; that was the nature of technology. "So the

question is, what risks are we talking about?" queried a biochemist who had worked in the industry for many years.

> Is there a risk of human safety? I mean to the extent that we have studied it, there is nothing that we can uncover that says that "this has risk." So at that point, you have to say, "You know what? I can't find any risk here, so yeah, it's risk free." But then, . . . is it possible that tomorrow somebody may have a new tool and a new way of looking at it? Possible, but not probable. So if anyone says to you, something is totally risk free, then they're inhaling stuff. There is no such thing as risk free.[26]

If scientists and inventors were held to too high a standard, another scientist suggested, no technologies would ever be accepted by society or make it onto the market.

> If you're of that belief that any potential for a hazard is reason not to market a product—and you know, I hear these adages all the time—well, there never would have been automobiles. There never would have been airplanes; most of the drugs that are on the market wouldn't be on the market. . . . We wouldn't be talking on the phone right now if everything that was a potential hazard never made it to the marketplace.[27]

The more important question had to do with the magnitude of those risks and how they stood up against the benefits these technologies offered. It thus came down to choosing the best alternative. What people need to understand, noted one plant biologist, "is that there are going to be payoffs down the line. It's risk assessment, [so] what do you want to do? Do you want to spray a hundred million gallons of pesticides on cotton fields in the South, or do you want transgenic cotton with *Bt* toxin?"[28] From many scientists' perspectives, the expectation that any technology should be risk free bordered on the ridiculous. This was certainly the case for the technology in which they and their companies had invested so heavily.

Doing Good by Doing Well

For the scientists who formed the research core of the emerging biotechnology industry, the 1980s were unique not only because people were able to do cutting-edge research backed by bulging budgets but also because the emerging toolkit of genetic technologies seemed to offer an opportunity to do something positive for the world beyond the contribution of producing new knowledge—and to get paid for it at the same time. "The discussion of the day was, 'we can save the world, we can feed the world, we can improve productivity, [and] farmers will make more money,'" recalled one industry scientist.[29] Identifying new drugs that could help people cope with health problems, discovering ways to produce pharmaceuticals more cheaply and efficiently than they had been in the past, finding better solutions to farmers' pest and weed problems— all these projects had social merit, and many scientists felt good about the possibility that their work might contribute something good to society.

Agricultural scientists enthusiastically embraced the new biotechnologies because they offered enormous potential to improve agricultural productivity. Often trained in U.S. land grant universities, these scientists took for granted the idea that improving agricultural productivity and addressing (mainly U.S.) farmers' problems were the ultimate goals of agricultural research. Genetic engineering was merely a way of augmenting yields and addressing the problems of agriculture more quickly and efficiently than had been possible using traditional methods of plant and animal breeding. Furthermore, in many scientists' minds, the new methods of gene transfer were more precise. "We were only putting one gene in, and we know exactly where it went," noted a specialist in plant virology.[30] "In the past we played roulette. We now have control over where the ball lands," noted another biotechnology researcher.[31]

As most bioscientists saw it, genetic engineering was simply one more in a long line of advances in the way human beings produce their food. "My basic premise is that genetic technology is simply a continuation of all other aggregate technology in agriculture," explained one scientist-turned-biotech entrepreneur.

> It started [long] ago [with] tractors, fertilizer, herbicides, insecticides, labor-saving devices, refrigeration, transportation, and new varieties, and in recent decades, genetic alterations using transgenic approaches. . . . I don't see that, from the corporate perspective, there was any difference at the time. This was [just] another way to improve productivity.[32]

Contributing to industry scientists' belief in the value of biotechnology were the environmental benefits they were convinced it would purvey. From their standpoint, genetic engineering offered a means of moving agriculture away from a heavy reliance on toxic chemicals and toward agricultural production practices that were more environmentally benign. "[We] believed in it," explained an entomologist who was motivated to work in industry by her desire to bring nonchemical pest control solutions to the market. "[We] were trying to get away from the toxic legacy.

I mean if . . . you have fifteen sprays of nasty chemicals on cotton, [or] you're putting in a *Bt* gene (*Bt* being very safe), then . . . it's a no-brainer."[33] According to this scientist, many of her colleagues were caught off guard by the social opposition to genetic engineering, precisely because they believed they were doing something beneficial for the environment.

> The scientists—well, the senior managers all the way down to the scientists—all had a very, very strong belief that we were doing something good for the world. And it was quite surprising for Monsanto that it would be seen as the opposite [of that], because everybody thought, "Here we're taking chemicals off the market and greening up the world with this new technology."

Such words reveal the way industry scientists understood genetic engineering and the societal role they saw for themselves. In their minds, they were doing good by doing well: reducing the environmental impacts of industrial agriculture, improving agricultural productivity, and addressing the needs of U.S. farmers. What is more, the technologies they were using to achieve these lofty goals were completely safe, at least according to available evidence.

Industry scientists' views about the promise of biotechnology and the good it could do for the world had a powerful impact on their business counterparts in the industry. Indeed, in the process of working together closely, these two groups became more aligned in their perspectives as the scientists shared their enthusiasm for and belief in the technology and the managers absorbed this positive energy and began to run with it. Over time, the managers of these corporations became biotechnology's most vocal supporters, combining their growing appreciation of genetic engineering with their keen understanding of what it took to introduce and market a new product. The fact that some of these businesspeople were scientists themselves, although they were not working at the bench anymore, only helped to solidify this shared perspective.[34]

Views of the Opposition

On May 10, 2004, the following exchange took place between a microbiologist from a small U.S.-based biotechnology firm and me, Rachel:

> INDUSTRY SCIENTIST: The bottom line in [our] small company, however, was that there ended up being very little "diversity of opinion" (at least in my opinion). There was a "party line," and that was that anybody who had a problem with ag genetic engineering (1) didn't understand it (i.e., didn't know what they were talking about), (2) was a Luddite, or (3) wasn't doing good science.

RS: Who created the party line, and how was it socially maintained? How did it affect the way employees interacted with the opposition?

INDUSTRY SCIENTIST: This is, of course, my interpretation of the situation. It's a conclusion I've come to in trying to make sense of what I felt was/is a widespread—that is, it went beyond [our company]—close-mindedness among scientists doing molecular biology–genetic engineering.

I don't know exactly who created the party line, but most of the most-respected, top-level plant molecular biologists I know of recite it. . . . My theory is that it relates to the reductionist way of looking at genes and their expression put forth by Jim Watson himself. I think it also relates to the fervent belief molecular biologists have in the potential of the technology; they get so caught up in the good they want to do that they just can't fathom that there could be a downside.

The result was that [my colleagues] and most other scientists utilizing molecular biology—whether industrial or academic—didn't believe the issues raised by opponents were valid.

As this exchange suggests, industry scientists' steadfast belief in the safety of the technology and its potential for doing good played an important role in shaping how these individuals viewed the critics of agricultural biotechnology and the arguments these critics made against genetic engineering writ large. From the typical industry scientist's, and businessperson's, perspective, anyone who organized against GMOs or was sympathetic to critics' claims that they were unsafe to eat or bad for the environment either lacked the scientific background to understand how the technology worked, was an anti-technology "neo-Luddite," or was motivated by political concerns. As a result, their criticisms were seen as invalid.[35]

Hearing industry scientists describe themselves, as well as the activists, puts these ideas into relief and reveals the immensity of the chasm that industry scientists saw between themselves and biotechnology's critics. In this regard, one plant biochemist's words are worth quoting at length. "Ours is a very rational culture," explained this former Monsanto scientist as he sought to make sense of what his opponents were thinking.

Well, the industry, industry scientists, we process everything through the lenses of "All right, what are the facts? Not what are the emotions, [but] what are the facts?" . . . So it is that cultural lens of trying to look at the world in such a rational, organized, systematic fashion and then process everything methodically and figure out the solution, which is what the industry [tries to do].

But to make the connection the way the activists can come at it . . .
"So cloning is bad." *"Why is cloning bad?"* "Cloning is bad because
my religious leader said so." *"Why did the religious leader say so?"*
"I don't know; you go to talk to him or her. God told me in my dreams
that I should say so." So how do you process that?![36]

In other words, whereas this scientist viewed the world from a factual,
scientific, and (therefore) rational perspective, anti-biotechnology activists and their followers were characterized as irrational, chaotic thinkers,
motivated by faith instead of fact. Another scientist, while speaking more
respectfully about those opposed to the technology, compared them to
his relative who was a Jehovah's Witness. In his view, it was impossible
to convince someone who believed deeply in something that things were
actually otherwise, no matter what the facts of the situation were. It did
not even pay to try.

The notion that the general public was uneducated about basic science
was a commonly held view among both the scientists and the businesspeople we interviewed, and it contributed to their belief that those who
opposed genetic engineering often did so out of ignorance. One biologist recounted a story about a British colleague whose wife had gone
to a supermarket to ask for a brand of genetically engineered tomato
paste that had been marketed in Britain earlier that year. According to
his colleague, the young woman at the counter looked at the man's wife
and asked her, "Do you eat those tomatoes with genes in them?" In his
opinion, that experience spoke volumes about the level of ignorance of
the average person. "That to me is the range," he concluded. "A company
that did everything right and [was] successful with the technology and
[was] making money, labeled the can, okay, and somebody at the other
extreme . . . does not understand the smallest bit of knowledge that she
eats . . . fifteen hundred meters of DNA every time she eats lunch!"[37]

Because most industry scientists and their business colleagues believed
that most activists did not understand even the fundamentals of science,
much less genetic engineering, it was easy to dismiss their arguments
and label the activists as charlatans.[38] In their minds, the activists simply
did not get it. "I mean, if you could grow corn under water stress, what
could be better than that?" asked one industry scientist and official.
"Why should there be any resentment or resistance to that technology? It
still blows my mind to this day, to be honest."[39] Adding to their disdain
was their belief that many activists purposefully manipulated the facts in
service of their own political views. "Whatever their [the activists'] support base is, whatever it is that they want to promote," reported a former

Monsanto scientist, "the fact of the matter is they don't mind twisting and turning and throwing stuff and connecting stuff in a way where even if it doesn't have a rationality, it still, however, is very appealing emotionally."[40] Another singled out the organic foods movement as a particularly despicable bunch because of their willingness to twist the truth in order to make money:

> Those who have taken the anti-technology [thing] farther, I think, are the organic group who buy [the anti-biotech activists'] claims, *claims only, no science!* "Don't take a chance; buy organic," [the activists say]. They can't claim more safety. They can't claim higher nutrition. They have no claims. They don't have anything. All they can say is, "Don't take a chance; buy us instead," and that's driven by money.[41]

Those who viewed genetic engineering as environmentally beneficial were perhaps the most aggravated with the activist community. From their perspective, biotechnology's critics completely missed the environmental benefits that gene splicing could offer, even though better environmental practices were supposedly what these critics were after. Industry actors found this apparent contradiction both maddening and unintelligible. Shaking his head in frustration, one plant biologist burst out:

> The science is really moving ahead and can deliver much more than it's been allowed to do, [for] the public good. There's some wonderful technology there. You know, you could develop strawberries that no longer have to be treated with fungicides to get rid of the gray mold on them. You remember how moldy and soft strawberries get within a few days when you take them home? You [could] stop all that. And you know that strawberries are bathed in fungicides before they put them to market.
>
> And here's the irony of this thing. Most of those are grown in California, . . . where the biggest amount [of pesticides are]—[and] where you have Mendocino County not allowing biotech crops [to be] grown in their county. Yet they are the ones who are suffering, who are getting the heaviest chemical load. Go figure! It just doesn't make any logical sense. . . . At the end of the day, you say, "Wait a minute, all we're trying to do is make a cleaner, healthier product!"[42]

Another example to which such environmentally minded scientists often referred was *Bacillus thuringiensis* crops. These were the crops into which insect resistance had been engineered, allowing farmers to reduce the amount of pesticides going into the environment. For them, using *Bt* crops should have been a no-brainer to the environmentally concerned.

But that was not how the activists saw it. Such thinking was incomprehensible to industry scientists and managers.

As the preceding discussion suggests, the lifeworld of industry scientists in the 1980s and 1990s had a certain insularity. One of the scientists we interviewed said as much herself. "They didn't go out a lot," she noted about her colleagues, contrasting them to herself, who went out frequently to speak to different audiences. "You know, I think a lot of them were just pretty much sheltered."[43] Most industry scientists saw themselves as highly knowledgeable and unbiased experts who knew the facts about the science and interpreted those facts according to norms of the profession. In their lifeworld, facts were facts, evidence was evidence, and context was irrelevant. They did not see theirs as a sociocultural world with its own norms of behavior, shared ideas, collective orientations, and assumptions. Worldview and politics had no place in this schema; indeed, they were totally irrelevant. All these views served to limit their sensitivity to and understanding of the opposition.

The Lifeworld of Industry

Industry scientists were not the only ones who brought with them a set of deeply held beliefs, "commonsense" ideas, and a certain way of looking at the world when they came to work in biotechnology. Corporate executives and managers who ran the companies in which these scientists were employed brought the same. The "business side" of the biotech industry comprised people who had generally spent years, often decades, working in industry and had typically received some kind of formal business school training as well. Many were veterans of the agriculture, chemical, and pharmaceutical industries and had worked for the same corporation for years. Both their informal on-the-job experiences and their formal business training had instilled in them certain ideas about what corporations like theirs needed to do to achieve success and how one managed to stay ahead of the competition. Starting in the 1980s, thanks to deregulation of the U.S. financial market, they began to learn another lesson, this one relating to the need to maximize short-term profits and what was coming to be known as "shareholder value."[44] In short, these business veterans came to the biotechnology business armed with ways of thinking and doing things that emerged out of their business experiences, both past and present.

At the most general level, of course, the business men and women running these large multinational corporations were oriented toward three broad and intimately related goals: staying profitable, growing the business, and beating the competition. No business executive had any doubt that his or her primary objective, and indeed the company's raison d'être, was to make money and to convey clearly to the market that the company had a strong growth trajectory. The health and value of their companies were continually being judged by Wall Street, which was well-known (and oft-repeated) in the offices, halls, and laboratories of these firms. Indeed, virtually all senior executives' compensation packages were heavily tied to Wall Street's evaluations. When the company's share value went up, so did the value of their (typically massive) stock options; when it went down, so did a sizable fraction of their fortunes. The investments these firms made in biotechnology, therefore, had one central end: to make a rising stream of profits for their companies.

Although all the big firms that sought to develop their businesses on the basis of biotechnology in the 1980s and 1990s operated with a clear and omnipresent market mentality, not all of them went about trying in the same way to achieve the goals of firm growth, profitability, and beating the competition. Some of the behaviors in which they engaged were shared broadly across the industry, whereas others varied significantly by company. In the case of common or shared behaviors, much can be attributed to the "large conglomerate" mode of operation and corporate mentality of the time, as well as to certain (emergent) characteristics of the biotechnology industry, such as its heavy reliance on intellectual property. In the case of major firm differences, individual firm cultures often played a powerful role.

The Modus Operandi of the Large Corporation

As noted earlier in this chapter, when companies such as Monsanto, Novartis, and DuPont decided to start doing research on molecular biotechnology, they invariably committed a large quantity of funds to the endeavor. One of the reasons these firms made such big investments was their sheer net worth, which reached into the tens of billions of dollars. The other reason, however, had to do with the fact that these firms were in science-based industries, in which the value of a company depended on its investments in R&D. Norms in the pharmaceutical industry, for instance, dictated that firms would spend some 15 percent of their revenues on R&D, and chemical and agrochemical companies typically allocated 10–12 percent of their revenues. For a twenty-five-billion-dollar company, even 10 percent represented a substantial sum of money.

As science-based enterprises, companies like Monsanto and DuPont were thus accustomed to investing immense resources in projects that could take years or even decades to bear any fruit. Furthermore, the most attractive kind of R&D investment for such a company was one that promised to yield major payoffs when the science finally did yield marketable commodities. The words of one long-time Monsanto veteran are worth quoting at length here, because they convey the kind of logic that many large companies applied when they considered making new R&D investments.

> I came up with a sequential series of hypotheses that [my colleague] and I . . . decided were valid to release to practice. Number one, Monsanto was quite used to accepting major risk in taking [new] technology [and] new science to the marketplace. The thing you have to accept when you're doing that sort of technology is that . . . you want to identify a problem that's extremely difficult to solve, [one] which is not so simple that, hell, two winos down the street can go solve it, and, you know, what have you brought to the party? So unless you get a problem, a need, that is extremely difficult to solve, then why waste all these tremendously expensive, talented resources to go work on [it]?
>
> Number two, it has to be extremely large, because . . . you can't afford to go find something for fun or sport, you know! You better pick out a big problem that will significantly impact agriculture.
>
> Number [three], it not only has to be a problem today; you've got to make a good enough bet that it's going to be problem ten, fifteen years from now, because it's going to take you that long to solve the problem. . . . You aren't going to say, "Eureka!" and run down the hall and say, "I got it!" in the first year or the second year or the fifth year. By and large, it's going to take you at least ten and probably twenty years. We started looking for what Roundup does in 1952, and we found it in '69, and we commercialized it in '75, some twenty-three years after we started looking. So it's a major commitment, a long-term commitment.
>
> And the other corollary to making a decision to do that type of research, [is that] the solution we find must do a very good job and we must be able to offer that solution to the farmer for a cost . . . such that we can make substantial profit using it, and he can make a profit. If the farmer and we both can't make profit, then we aren't going anywhere.
>
> And finally it's got to be a long-term solution, because if the farmer will only use our product for a couple years, then . . . we find out something's bad about it, you know [that is no good]. We've got to be in that marketplace for a long time.[45]

Of course, investing large sums of money in research that might not yield any revenues for ten to twenty years carried a serious risk: if these investment decisions did not turn out to be sound, they could create tremendous losses for a company. This looming reality put pressure on firms to do all they could to ensure that their R&D investments paid off handsomely.

When these major multinationals decided to reinvent themselves as life science companies, they carried out their plans using the familiar tools of large-scale, corporate restructuring: they sold off those businesses that served to detract from their new corporate image and desired areas of operations and bought up companies that advanced their strategic plans. Acquiring specialized biotechnology start-ups that had the scientific expertise and intellectual property they wanted became one central element of their strategy; vertically integrating into the seed sector became another. Gaining access to the seed business was important, because without access to seeds, or what scientists refer to as "germplasm"—the set of genes that shape the characteristics of each organism—firms could not get their genes out of the lab and into farmers' hands, that is, into the marketplace. By investing in companies that possessed the germplasm, infrastructure, and reputation for producing and distributing large quantities of seed, agricultural biotechnology companies could solve this market access problem. Many MNCs also engaged in horizontal integration: large pharmaceutical firms, for example, sought to acquire new subsidiaries that specialized in agriculture and nutrition, and agrochemical companies sought to acquire or merge themselves with drug companies.

A brief look at the history of one of these companies, Ciba-Geigy, offers insight into the corporate machinations involved in this process, as well as into the size of the enterprise that emerged. Ciba-Geigy was the product of a 1971 merger between two century-old chemical companies based in Basel, Switzerland.[46] In 1974, in an effort to complement its businesses in agrochemicals, Ciba-Geigy expanded into the seed industry with the purchase of a U.S.-based seed company called Funk Seeds. It then advanced into biotechnology in 1980 with the establishment of a special biotechnology research unit. Absent a strong presence in the medical biotechnology area, the company formed a strategic partnership with a promising California-based medical biotechnology company named Chiron in 1994.[47] Just two years later, Ciba-Geigy decided to merge with another industrial giant, Sandoz, whose primary strength was in generic drug production, although it also had interests in nutrition and agribusiness. Among the many assets that Sandoz brought to the deal were three major seed companies. Going under the new name Novartis, Ciba-Geigy and Sandoz's marriage was heralded as "one of the largest corporate mergers in history."[48] In the year of the merger, Novartis was worth eighty billion dollars and had over 116,000 employees working in close to one hundred countries around the world.[49]

By the middle of the 1990s, most of the companies that decided to follow a life science model looked fairly similar to Novartis; they had investments in agricultural biotechnology, seed companies, agrochemicals, pharmaceuticals, and nutrition. Perhaps even more important, though, was the collective impact of their restructuring efforts on the structures of several of these other industries. In the process of doing "business as usual," these large conglomerates had collectively acquired the most important biotechnology and seed companies on the market. The process of industry consolidation was well underway (see Table 1).

Making Biotechnology into a Profitable Business

As noted above, when firms such as Monsanto, DuPont, and Ciba-Geigy (later Novartis) moved into the biotechnology area, their executives came with some very clear ideas about what they had to do to make biotechnology into a profitable business. One of these had to do with the need to establish property rights. These executives came from industries that were heavily dependent on intellectual property protection, as the pharmaceutical and agrochemical industries were, so the need to have property rights over scientific discoveries was a standard element of their business strategies. In effect, competing for patent rights over genes and gene transformations became a "first principle" of the business.

Recognition of the need for property rights was not limited to those who represented the "corporate side" of the biotech business; it extended to the scientific staffs of these corporations as well. Industry scientists came to understand intellectual property (or "IP," as it is called in the vernacular) as a necessary aspect of the business, one that their companies needed to focus on if they were going to succeed in the marketplace, and one that they as individuals needed to focus on if they were going to maintain research freedom in their labs.[50] As one scientist put it, "Most of the value you get out of them is, you get freedom to operate. You get freedom to continue using something you discovered. . . . [Otherwise], somebody else is going to make the same discovery, patent it, and then sue you to make you stop."[51] Another industry scientist explained the necessity of patents to agricultural biotechnology firms:

Table 1. Estimated seed sales and shares of U.S. market for major field crops, 1997

Company	Total sales (in millions of dollars)	Total market share (percent)	Corn market share (percent)	Soybean market share (percent)	Cotton market share (percent)
Pioneer Hi-Bred	$1,178	33.6	42	19	0
Monsanto	541	15.4	14	19	11
Novartis	262	7.5	9	5	0
Delta & Pine Land	79	2.3	0	0	73
Dow Agrosciences/ Mycogen	136	3.9	4	4	0
Golden Harvest	93	2.6	4	0	0
AgrEvo/Cargill	93	2.6	4	0	0
Others	1,121	32.0	23	53	16

Notes: Total market shares in this table include only corn, soybeans, and cotton. Monsanto acquired DEKALB in 1997 and Asgrow in 1998; the planned Monsanto and Delta & Pine Land merger was called off in December 1999. Columns may not add to the percentages shown because of rounding.

Source: Fernandez-Cornejo 2004, 27 (Table 13).

> You *have* to have them . . . because it's a regulated industry and because
> of the heavy investment in R&D. It's the same in the pharmaceuticals
> and the high-tech industry. . . . If you're in a regulated industry and you
> have such a long, long lead time and huge R&D costs, you have to get
> a return on that. And there's not enough money in it if . . . there is no
> intellectual property and everybody shares, and everybody gets a little
> piece of a big pie instead of a big hunk of the big pie.[52]

Indeed, patents were so normalized in the industry that no one ever really
stopped to think about them. As we will see in chapter 3, however, what
seemed absolutely necessary and unquestionable for these businesspeople
and industry scientists—something "you had to have"—was a question-
able assumption for others.

A second idea that made "common sense" to the business staffs of
these companies (and one that they constantly sought to impress upon
the scientists in their midst) related to the kind of products their companies should focus on developing. Not surprisingly, the products of greatest interest were those that offered the largest market potential. Market
potential was typically defined in terms of sales volume (and in the case
of agricultural biotechnology, in terms of acreage application), but it also
involved an element of market longevity, as the Monsanto veteran's earlier example of the development of Roundup suggests. A market focus
meant that while some sorts of research were highly appealing to the
executives of these large companies, others were considered a waste of
time because they did not anticipate sufficient demand. It was this basic
business reality that explained why companies aggressively sought to develop herbicide-resistant plants and crops into which the naturally occurring insecticide *Bacillus thuringiensis* could be engineered, and why
they generally avoided pursuing others that might have had more value
from a societal perspective, such as nutritionally enhanced cereal crops
and drought-resistant crops cultivated mainly by farmers in the global
South. Herbicide resistance was attractive not only because it could be
engineered into crops such as corn, soy, and cotton, which accounted for
hundreds of millions of acres of U.S. farm production, but also because it
worked in conjunction with proprietary herbicides, such as Monsanto's
Roundup and AgrEvo's Liberty. Thus, companies had the opportunity to
make money both on the genetically engineered seeds *and* on the herbicides that went with them.

The logic large companies applied became clear in a discussion we had
with a regulatory scientist at one of the top agricultural biotechnology
companies in the industry. When asked how his company decided which

products to pursue and whether this logic varied much among firms in the industry, he answered:

> I think we're all very alike. Let me tell you an almost hypothetical story. Let's say, through either reading the literature or something, I know that a particular organism produces a compound that would be effective in controlling . . . mycotoxins in wheat. That's a big issue, right? Millions and millions of tons of wheat are thrown away every year because they're riddled with mycotoxins. So this would be a really good thing if we could have a wheat plant genetically engineered [such that] mycotoxins would not accumulate. And so I'm a scientist here, and I'm walking in to my management saying, "I have a gene that will control mycotoxins in wheat, which will be fantastic."
>
> OK. The first thing that [my company] will do is, they'll ask me to put together a research plan. In other words, how many years? How many people? How much money is it going to cost? And then they're going to turn to the business guys and say, "Tell me how much money I'm going to make in year one, year two, year five, and year ten." And the decision about whether that research project is going to go forward, whether it's a good one or bad one, will be [based on] how much money I am going to spend as a scientist versus the MBA telling them how much they're going to be able to make. And if those ratios aren't good enough, that project never gets out.[53]

In short, if he as an industry scientist could not come up with an idea that would make enough money for his company, then his company was not going to be interested.[54] On the other hand, he explained, "If the company says yes to your project, you have all the money you need to try and get it done." Aside from reflecting the extent to which firm scientists had absorbed this kind of economic logic, this anecdote reveals the power of the incentive system involved.

A third set of ideas the "corporate side" of the business brought to the table related to notions about how to compete in the marketplace. These ideas derived largely from these individuals' experiences working in science-based companies and in industries that were characterized by a situation of "oligopoly," in which relatively few firms make and sell the same goods.[55] Generally speaking, this set of ideas boiled down to three closely intertwined principles, one of which we have already mentioned, namely, the establishment of intellectual property. Establishing IP was important because it allowed a firm to carve out its "technology position" in the marketplace, meaning it owned the technology that was in high demand. The second principle of competition most business executives embraced was the need for a firm to get its products into the market ahead of its competitors.[56] The advantage of being first was that it gave a firm a head start in building product familiarity and loyalty. Once customers had tried the firm's product and discovered they liked it, they would be reluctant to try another, since doing so would require them to assume some risk. Being "first to market" offered another important benefit as well: it typically raised the value of a company's stock, especially if the product was the first in its particular class. The third principle of competition involved competing for market share. The industries in which these large conglomerates were most active (i.e., pharmaceuticals, chemicals, agrochemicals, energy) are all oligopolies. Competition is often extremely fierce in an oligopoly, but rather than focusing on eliminating the competition, an oligopolistic firm coexists with its rivals in a contested terrain where companies struggle for dominance but do not expect to exert total control. "Owning the market" thus translates to holding the largest share.

In sum, all of the conglomerates that sought to convert themselves into life science companies in the 1980s and 1990s followed the broad approach just outlined, for roughly similar reasons. They bought and sold off parts of the business using familiar tools of corporate restructuring, they invested in projects that promised to capture them the largest markets and best stock value appraisal, and they sought to create a portfolio of intellectual property with which they could establish a strong "technology position." Nonetheless, even within this world of similarity, important differences occurred in the commitment these companies made to biotechnology and the strategies they employed once they entered the business. Of all the firms that moved into agricultural biotechnology, Monsanto made the greatest commitment, allocating the most resources to this vision and most faithfully "staying the course" in order to get there. By the mid-1980s, this long-time chemical company from St. Louis had emerged as the industry's leader, a position it would hold for the next twenty-five years.[57] As industry leader, Monsanto set the tone for much of the industry with its corporate decisions, growth strategy, and way of doing business. It was also the face that the activists (and the rest of the world) most often saw when they came into contact with the industry.

Corporate Culture and Firm Strategy: The Case of Monsanto

In her insightful book, *The Cultural Crisis of the Firm*, Erica Schoenberger argues that corporate culture and firm behavior, particularly firm strategy, are "mutually constitutive," meaning that corporate culture both reflects the past business strategies of firms and helps produce the strategies they develop to lead them into the future. Schoenberger goes on to make the point that "corporations are run by real people," and "to understand corporate strategies, we need to understand something about corporate strategists. Specifically, we need to consider what shapes their interpretations of the world and their ability to act on it."[58] In the following discussion, we describe some of the historical experiences that shaped Monsanto's strategies and its *strategists,* that is, the key decision makers in the firm. Such an excursus can help us understand how the company cognitively and materially constructed the world of biotechnology and how that construction, in turn, generated certain behaviors that fueled and antagonized the opposition.

A History in Agrochemicals

As early as 1975, John W. Hanley, Monsanto's CEO at the time, began positioning the company to become a world leader in genetic engineering. A former executive at Procter & Gamble and an MBA from the Harvard Business School, Hanley was hired by Monsanto because he was opinionated, decisive, and a "take charge" kind of guy. The first outsider to take the reins of this seventy-five-year-old, deeply midwestern company, Hanley was committed to professionalizing what he saw as a highly personalized and idiosyncratic investment culture. He was also extremely enthusiastic about biotechnology.[59]

In the years leading up to and following Hanley's decision to shift the company toward a future in biotechnology, Monsanto had arguably become the leading agrochemical company in the country, if not the world. From being ranked around sixtieth in terms of agrochemical sales in the 1950s, it had climbed into the top position by the 1980s. Much of the reason for the company's rise had to do with its mutually reinforcing internal culture and business strategy. Anxious to improve its position in the agrochemical industry, Monsanto worked hard to develop several new proprietary herbicides and insecticides, including the blockbuster herbicide Roundup. As each new product was developed, Monsanto aggressively marketed it to farmers, steadfastly focusing on the goal of capturing market share. To accomplish this objective, Monsanto developed a dense network of relationships with farmers, farmers' cooperatives, and other local suppliers of agrochemicals and used promotions, economic incentives, and advertising to acquaint its customers with and generate their loyalty to its products. Monsanto's marketing network reached deep into the U.S. farm community, particularly in the Midwest, where its main agrochemical markets were based and where the company's headquarters was located. If there was one thing this company knew and knew well, it was midwestern farmers.

The close contact Monsanto established with the U.S. farm community served both as a vehicle for getting its products into the hands of customers and as an information conduit for the company's research pipeline. More than most of its competitors in the agrochemical business, Monsanto was "market-driven," meaning that it sought to identify the problems its customers faced and then went back to the lab to try to solve them. "It was always a very scientific, market-driven [company]," reported one former company employee. "Every one of us was required to go out and look at all the field trials of all the new products, from the CEO on down to a certain level in management, [to] see for ourselves what the new products looked like in the field."[60]

Monsanto's highly aggressive business culture complemented its strong market orientation and manifested itself in myriad ways. Corporate culture at Monsanto was, as one former manager put it, "dog-eat-dog." "I had to develop extremely broad shoulders and a really tough skin," this scientist–manager reported. "I was successful because I was able to do that. But [it was] highly political. There was always jockeying for who was going to get promoted."[61] When it came to interactions with the outside world, company culture was equally aggressive. Monsanto pushed its marketing employees to achieve high product turnover among its customer base. "[It was] extremely pushy for results . . . pressing the customers, to get the products out," noted an employee who had recently retired.[62] Monsanto also took a forceful stance with respect to other companies with which it wanted to do business, seeking to gain and maintain the upper hand in these relationships. The same individual just quoted offered a revealing insight into Monsanto's "it's all about us" attitude:

> Internally, Monsanto, when they developed strategy, they would always talk about "value capturing," capturing the value of that gene through the chain, which was fine internally in the analysis ("how much of that ten dollars at consumer level can we capture?"). And you do that one, or two, or three years while the scientists work it out. [But] then . . . the sales rep goes to, say, a Pfizer, and says, "Here we have this superb health benefit in seed, and . . . our value-capturing strategy is the following. . . ."
>
> And Pfizer says, "What do you mean? It's *our* product, it's *our* market. *We* are the pharmaceutical here. What are you talking about

capturing, Monsanto? Are you going to come do our job or what?" So instead of using simply the word *value sharing*, . . . the internal jargon of *value capturing* . . . when you transpose it outside in negotiation, suddenly becomes extremely acidic, extremely rough, for the people you talk to.

As a result of this attitude, many of their business partners perceived the company as arrogant, he suggested.

The company's success in the agrochemical sector also stemmed from the close relationships it developed with government regulatory agencies and agricultural officials. From early on, Monsanto recognized the importance of paving the path with government lawmakers and regulators in order to ensure that its products would make it to the market in a timely fashion. Gaining government regulatory approval offered another invaluable benefit in the minds of Monsanto executives: this approval said to the public that these products were safe and had been validated as such by the U.S. government.[63]

As its concern with regulatory affairs suggests, Monsanto was a firm that was very much about planning. Years before Monsanto's popular herbicide of the 1970s, called Lasso, was scheduled to go off patent, the company began searching for a new herbicide that could kill an even broader spectrum of weeds. The product its scientists eventually developed was Roundup, which turned out to be a blockbuster. Similarly, years before Roundup was ready to meet the same fate, Monsanto executives were back in the meeting room, planning for the day when Roundup would no longer be protected under U.S. patent law. One former employee described Monsanto's planning culture with admiration: "It was a big company. They'd plan; we'd go off on a retreat, and they would show us what . . . the market planning would do. And they planned for the expiration of the Roundup patent ten years before it happened. It was remarkable!" she exclaimed.[64]

Monsanto's success in agrochemicals and its position as an industry leader had a powerful influence on the way its managers and employees perceived themselves in the business world. Those who worked at Monsanto were proud of working for such a successful company and of the contributions that they themselves had made to that success.[65] Monsanto employees had a sense of themselves as competent, knowledgeable, and successful businesspeople. They knew, and *firmly believed they knew*, every dimension of the business. Of course, from an outsider's perspective, such supreme confidence could easily be read as arrogance.

Monsanto's Move into Biotechnology

When Monsanto began investing in the new biosciences, it imported many of its ideas, beliefs and behaviors into the business of biotechnology. One of the arenas in which Monsanto's corporate culture made its imprint was the way the company went about building up this new element of its business. Unlike the other large conglomerates, which invested slowly and cautiously in biotechnology, Monsanto strove to become the industry leader from the beginning and stayed focused on that goal for the next thirty years. As a scientist who worked for one of Monsanto's competitors observed,

> [Monsanto executives] basically took off their gloves years ago and completely focused all of their energies in succeeding in biotechnology. DuPont, Dow, Syngenta . . . all of those companies took a much more cautious approach. . . . Syngenta and Dow and others said, "Well, we're going to do biotech as one component of our company but still rely on the traditional part of our company to drive most of the profits." Monsanto said, "We're going to make all of our profits from biotech." And they succeeded![66]

Indeed, while other companies often dramatically changed their business plans when a new CEO took over the operation, this did not happen at Monsanto after 1975. Three successive chief executive officers—John Hanley (who retired in 1984), Richard Mahoney (1984–95), and Robert Shapiro (1996–2000)—all followed the same course, channeling an ever-increasing share of the company's resources into biotechnology and moving Monsanto away from its chemical-industry past. Shapiro was the most taken with the life science idea, betting the company's future on biotechnology in the belief that it could make the company piles of money while creating a more environmentally sustainable world.

Once a major commitment to biotechnology had been made at the highest levels of the corporation, Monsanto's scientists, business strategists, regulatory affairs personnel, and product managers all kicked into high gear. These employees brought an immense amount of energy and focus to their jobs, buoyed by an influx of money and staff and motivated by a combination of competitive spirit and the economic urgency that accompanied the task of developing marketable products from an immature science. With hundreds of millions of dollars invested in the biosciences and little to show for it by the early 1990s, the company made it clear to its scientists that they needed to make this project work—and work soon. As then CEO Mahoney told them, "We are not in the business of the pursuit of knowledge; we are in the business of products."[67]

Reflecting their own sense of excitement as well as th/ sure from company management, Monsanto's bioscienti lessly to develop products that could be sold to the farm community. "There were three companies that spent the same amount of money on agricultural biotechnology, DuPont, Novartis, and Monsanto," noted Roger Malkin, the chairman of Delta and Pine Land Company, a large southern seed company with which Monsanto did business.

> And Monsanto is the only company to make anything of it. . . . The difference is all attitude. At Novartis there was no sense of urgency. If you went by their parking lot in Research Triangle Park on a Saturday, there were no cars in the parking lot. If you went to DuPont, those researchers all went home at 5:00 in the afternoon. At Monsanto there were always cars in the parking lot at 1:00 A.M. and on the weekends.[68]

Monsanto identified three sets of stakeholders that needed to be on-board in order for it to become the industry's leader. The first was farmers, or more specifically U.S. farmers. A focus on American farmers seemed the obvious strategy for the company, because they were the buyers of the genetically modified seeds Monsanto was hoping to sell. They were also the customer base with which Monsanto's Agricultural Division was most accustomed to working. As noted above, in the course of promoting its agricultural chemicals, particularly Roundup, Monsanto had erected an extensive marketing infrastructure that ran through every rural road in the countryside. As an agricultural executive for the company explained, U.S. food consumers were simply too far away from the food producers for the company to be that concerned about them.[69] In fact, it was unlikely that U.S. consumers would even know or care about the type of seeds being used to produce their food. Rather, Monsanto assumed that once it convinced farmers that its GM seeds were superior to other seeds, the bulk of its job was done, at least in terms of the market. As we will see in the next chapter, however, this assumption turned out to be false.

The second set of stakeholders Monsanto concerned itself with was government regulators. Government regulatory agencies were the ultimate gatekeepers to the market, and Monsanto saw them as crucial to establishing the company's "freedom to operate," or its right to introduce and commercialize its products around the world. Without government approval, Monsanto could not legally market its products, no matter how good and effective they might be. As a result, Monsanto allocated ormous resources to its regulatory affairs department. Indeed, long before any products came out of the lab, company officials were already working closely with the U.S. government and others to develop "acceptable" systems of biotechnology regulation.[70]

In this way, Monsanto stood apart from some of its competitors in the industry. Monsanto was not opposed to regulation but saw it as necessary for the company's successful operation. Regulation would protect the firm against future liability claims and instill consumer confidence in its products. Monsanto thus poured a lot of energy into achieving a favorable regulatory environment for its products, both at home and abroad.[71] So seriously did the company take the regulatory issue that it outspent every other company in the agricultural biotechnology industry in this area.[72] According to a public relations specialist at the company, Monsanto was considered to have one of the best government affairs departments in the industry in terms of "dealing with the regulators, getting things approved, and, you know, dealing with those people in a very effective way."[73] Its competitors apparently agreed. In the words of an official from another large biotech company, "Monsanto said, 'What do we need to do to win? You need to have unapproachable science. It has to be done very quickly, the regulatory science. And you need to gain approvals worldwide in order to market commodity products.' And they hired everybody they needed to, and then some, to get the job done worldwide. No one else has done that," he observed with admiration.[74]

The final groups Monsanto identified as key stakeholders included major shareholders and Wall Street more generally, without whose approval the company could not succeed. Virtually all of the company's employees, from the CEO on down, attuned themselves closely to the evaluations emanating from Wall Street, where every move the company made was carefully scrutinized and critically assessed by financial analysts from outside the firm. Monsanto officials thus focused their attention on establishing patents and getting their products into the market first, to make sure that the company would appear as the scientific leader in the industry. Monsanto managers assumed that if the company managed to achieve these goals, the rest would fall into place.

The Shapiro Era

When Robert Shapiro took over as Monsanto's chief executive officer in 1996, he carried the company's already serious commitment to biotechnology to a whole new level, both psychologically and financially. Shapiro was by all accounts a persuasive, inspiring, and motivational leader;

indeed, Monsanto employees described him as a "visionary" who swept people up with his larger sense of purpose and broad perspective on the technology. "[Bob was] very insightful, bright, sophisticated, so when he said, 'Let's run,' everybody said, 'We are going to run,'" noted a senior scientist in the company, who later went to work for DuPont.

> He had the whole enterprise mobilized behind him, and so if you walked the halls of Monsanto, people were going to work with a sense of purpose, feeling like owners, you know, not that they were a part of an organization drawing a paycheck, but they were there to make a difference. I mean that was down into the level of janitors to some extent. . . . That kind of infectious enthusiasm you don't see in a company like DuPont.[75]

Shapiro deeply believed that biotechnology was essential for achieving sustainable development and saw the life sciences as a vehicle for moving the world away from consuming more and more "stuff" and toward consuming and using knowledge.[76] The power of these ideas was not lost on company employees. As one employee described it, "There was a term around Monsanto, *Bobalooey,* because it almost became liberating," she recalled.

> He [Shapiro] said, "I don't want to reward people with carrots or sticks; I want people to reward themselves. I want people to be motivated themselves, out of purpose. . . . He and others started talking just so much about how our technologies could make this world better for ourselves and for our children. And we believed in the benefits of Monsanto's products so greatly that [we] became zealots almost, to get them to market. Here we were talking about "food, health, and hope" and all the benefits, and people thought we were crazy. But we became so inspired and purposeful, almost spiritual, about Monsanto and our products and the difference that we were going to make. . . . If you walk in that company today and you talk to people, they are very, very motivated, because we believe these products are going to make the world a better place.[77]

Shapiro's effect on the core businesses of the company was equally dramatic. He sold off the last of the firm's chemical concerns and aggressively sought to secure access to seed companies in order to ensure that the company could build market dominance in the field as well as in the laboratory. Between 1996 and 1998, Shapiro spent eight billion dollars to acquire half a dozen seed companies, including Asgrow Agronomics, a global leader in soybean research and seeds; Holden's Foundation Seeds, supplier to over a third of the U.S. corn market; Sementes Agroceres, a leading Brazilian corn seed company; Cargill's international and seed distribution operations; Plant Breeding International; and DEKALB Genetics.[78] As he explained his strategy to the company's stockholders,

> [We] knew that gene discovery alone wouldn't be enough to build a successful business. Farmers don't buy individual genes. They buy seed that contains the package of traits they want, and they want both traditional traits from conventional breeding and new ones from biotechnology. To succeed, we needed to ensure that farmers could buy our traits in the seed they want—crop by crop, region by region, country by country.
>
> We also believed that in this business speed would be critical: getting to market ahead of our competitors, with the right traits in the right germplasm in each key crop and market. This meant that, in addition to having the leading gene discovery programs, we would need to collaborate seamlessly with a large number of seed companies to get the crop and market coverage we needed to beat the competition to market. We explored a number of possible approaches . . . and concluded that combining a range of seed capabilities with our gene discovery capabilities into a single company would give us competitive advantages in speed and cost that would otherwise be unattainable.
>
> That's why we've been buying seed companies around the world. In a short time, we've pulled together the germplasm, the global market presence, and the talented, energetic people to help us bring our genes to market.[79]

In putting together such a comprehensive strategy for bringing their genes to market by buying up seed companies around the world, Shapiro made Monsanto appear as if it were trying to dominate world agriculture by gaining control of one of its most essential inputs: seeds. That interpretation was not so off-base, according to those who knew Shapiro and his company well. "Shapiro was driven, the leadership at Monsanto was driven, by this vision of the grand unifying life science company that would become an end-all and be-all, and be, you know, essentially the Microsoft of biosciences," a speechwriter for Shapiro later reflected.[80] Tom Urban, the former CEO of Pioneer Hi-Bred, one of the oldest, most respected and powerful seed companies in the United States, concurred: "Shapiro has this messianic sense about him. . . . If he said it once, he said it three or four times: Put us together and we'll rule the world. We're going to own the industry. Almost those exact words. We can be a juggernaut. Invincible."[81] While this strategy may have made perfect sense to those who worked in the industry, it was neither commonsensical nor acceptable to many of those who did not. Among them were a growing number of the technology's, and the company's, critics.

The Public Face of the Biotech Industry

As we have sought to show in this chapter, the scientists and business-people in the large multinational companies who came to dominate the agricultural biotechnology industry influenced one another's thinking and behavior in important ways. Working in a corporate setting, former university scientists learned to accept the parameters of business and to let industry managers' historically and culturally shaped understandings of how to make money from this new science define the kinds of projects on which they would work and the ways in which their knowledge and skills would be put to use. In exchange, these scientists enjoyed substantial salaries, unparalleled access to resources for their scientific work, considerable scientific freedom, and the opportunity to work with first-rate colleagues. For their part, the people who managed the industry were powerfully influenced by scientists' enthusiasm for genetic engineering and their views about biotechnology's superiority over existing technologies. Many corporate executives were excited about molecular biology's economic potential, coming at a time when, at least for the chemical industry, the future looked dim. Scientists' and corporate managers' positive outlooks on biotechnology mutually reinforced each other and helped generate the high levels of energy that drove industry scientists to make new discoveries and find applications that could become marketable commodities. At the same time, this scientific–corporate juggernaut existed in a relatively insulated sociocultural world. From within this lifeworld, it was hard to see—or at least take seriously—alternative and more critical interpretations of what it meant to engineer the genetic code purposefully in the name of economic profits.

Furthermore, as the firms that formed the core of the biotechnology industry acted in "normal" ways to innovate through science and to establish themselves as dynamic biotechnology companies, the biotechnology industry became highly consolidated. Leading the pack in the agricultural arena was the former agrochemical giant, Monsanto, whose behavior as frontrunner set the tone for the industry. By making certain strategic decisions and acting in ways that reflected its corporate culture, Monsanto executives and employees generated a public face for the industry and created rules of competition that other companies would have to take into account, if not abide by entirely. The public face the company created took the form of an arrogant and aggressive corporate actor whose goal was to push these new technologies as hard as it could, both to government regulators and in the marketplace. The rules of the game Monsanto helped to establish involved building the source of profit making directly into the seed through proprietary genetic manipulations and then fighting for market share by purchasing a critical complement of significant seed companies. Through its actions, the company created an image of an industry that was attempting to gain control of global agriculture through its ownership of a growing share of agricultural inputs. The most critical of those was the source of much of the world's food supply: the seed.

From the perspective of two very different sets of observers, Monsanto's behavior was highly problematic. Among its industry competitors, the company was seen as trying to dominate the market in genetically engineered crops and was anything but a team player. Some industry officials believed that Monsanto alone wanted to determine how the technology would be introduced and regulated and did not much care what other companies thought about these sensitive issues. Many also felt that in pushing so forcefully to get the technology into the marketplace, particularly in western Europe, the company had spoiled things for everybody. It had given the industry a bad name and created a consumer revolt in Europe that negatively affected the entire industry. A small but growing group of activists, who are the subject of the next chapter, were even more outraged by Monsanto's actions. From their vantage point, the agricultural biotechnology industry, led by this aggressive U.S. company, was guilty of trying to dominate world agriculture, including disenfranchising farmers from their very access to seeds. And that was only one element of their offensive behavior.

CHAPTER THREE

Forging a Global Movement

On March 7, 1987, thirty-one people from twenty-two countries gathered in the small village of Bogève, France, for a workshop titled "The Socioeconomic Impact of New Biotechnologies on Basic Health and Agriculture in the Third World." The Dag Hammarskjöld Foundation, based in Uppsala, Sweden, sponsored and funded the workshop, and the Rural Advancement Fund International (RAFI) organized it. Attending were members of a number of international NGOs: the International Organization of Consumers Unions, based in Penang, Malaysia; the Seeds campaign of the International Coalition for Development Action, in Barcelona, Spain; Health Action International, a network of organizations working toward "health for all" and in opposition to abusive practices by the pharmaceutical industry; the International Baby Food Action Network, a coalition of organizations that instituted the "No to Infant Formula" campaign against the Nestlé Corporation; the Pesticide Action Network, a set of regionally based organizations devoted to halting the use of dangerous chemicals in agriculture; and the Seeds Action Network, a network of groups that aimed to protect crop biodiversity around the world. NGO representatives, a handful of academics, and others from India, Peru, Brazil, the Philippines, Ethiopia, and the United States also attended. For four days, these individuals explained, discussed, and debated the "new genetics" and the related issues of genetic erosion and the new trends in patent law. Much of the conversation focused on the implications of biotechnology and these other phenomena for the third world, or what is now called the global South.[1]

By the end of the workshop, participants had collectively produced a powerful public statement outlining their position on biotechnology. Titled

"The Bogève Declaration: Towards a People-Oriented Biotechnology," most of it emphasized the potential hazards of biotechnology: "Biotechnology is a global issue. . . . Like any other technology, it is inextricably linked to the society in which it is created and used, and will be as socially just or unjust as its milieu. Therefore, we conclude that in today's world this most powerful new technology is more likely to serve the interests of the rich and powerful than the needs of the poor and powerless." While biotechnology offered an opportunity to improve the quality of life for humanity, the declaration suggested, in the current context such an outcome was a remote possibility. More probable was that biotechnology would have serious health, socioeconomic, and environmental consequences, some of which would be impossible to reverse. In agriculture in particular, genetic engineering "is . . . likely to accentuate inequalities in the farm population, aggravate the problem of genetic erosion and uniformity, undermine life-support systems, increase the vulnerability and dependence of farmers, and further concentrate the power of transnational agribusiness." In the health area, pharmaceutical companies could be expected to focus on the most profitable investments, drawing attention away from basic health needs.[2]

Such a bleak and negative assessment of biotechnology could not have been further from the hope and promise that many scientists and industry participants saw in genetic engineering. How could these two groups arrive at such tremendously divergent interpretations of a technology that in 1987 had barely made it out of the lab and into society? Why were the participants in the Bogève workshop so pessimistic about these new genetic technologies, even though the participants came from such different backgrounds and worked on very different sorts of issues? How did they arrive at such a dire view?

In this chapter, we focus on the people and ideas that formed the core of the anti-biotech movement, seeking to understand who these individuals were, how they developed a critical perspective on genetic engineering, and what motivated their efforts to organize around this issue. On the basis of a combination of in-depth interviews and historical–archival work, we show that the anti-biotech movement emerged out of a particular historical moment and a small group of people from the global North and South who had been politicized by their experiences with the Vietnam War, their participation in various social movements and causes of the 1960s and 1970s, and their exposure to critical ideas and ways of thinking. These people came to the issue from the development community, the environmental movement, and the scientific community.

Working together across tables and continents, telephone lines, and later the Internet, they developed a collective interpretation of genetic engineering—in effect, a grievance—that was diametrically opposed to the one expounded by the biotech industry. Like the view of the people who worked in the biotech industry, however, these critics' views of the technology were deeply rooted in their personal biographies, their world-views and values, and their locations in specific social networks—in other words, in their lifeworlds. As they brought their worldviews and moral sensibilities to bear on the questions surrounding biotechnology, they came to a very different conclusion from that of their industry counter-parts about how this technology could and would be used.

While all social movements embrace some criticism of the status quo, not every critical analysis that is articulated in society becomes the basis for a movement. Much depends on the ability of activists to translate their griev-ances into effective and sustained political engagement with their adversar-ies. Thus, a final question we address in this chapter is: How did this small group of critics blossom into a global social movement capable of negatively affecting the industry's fortunes and foiling some of its best-laid plans?[3]

Critical Communities and "Thinking Work"

In analyzing the process by which activists developed their countercultural analysis of biotechnology, we build upon some ideas from the political scientist Thomas Rochon. In his book *Culture Moves: Ideas, Activism, and Changing Values,* Rochon asks how social movements develop ideas that differ from dominant ways of thinking. According to Rochon, small groups of intellectuals, networked in "critical communities" of ideologi-cal production, play a central role in creating new theories and ways of thinking. These are people "whose experiences, reading, and interaction with each other help them to develop a set of cultural values that is out of step with the larger society."[4] Critical communities serve to provide new value orientations to particular issues, and they develop new discourses for apprehending them. Rather than simply thinking new thoughts within existing frameworks of interpretation, they fundamentally "alter the con-ceptual categories with which we give meaning to reality."[5] Thus, critical communities form a "countercultural current" within society.

Rochon's critical communities represent what we could call a *move-ment intelligentsia,* and his observations about how such communities form and what they do are apt for describing the process of grievance formation in the early phase of many social movements:[6]

The creation of new ideas occurs initially within a relatively small community of critical thinkers who have developed a sensitivity to some problem, an analysis of the sources of the problem, and a prescription for what should be done about the problem. These critical thinkers do not necessarily belong to a formally constituted organization, but they are part of a self-aware, mutually interacting group.[7]

Rochon notes that members of a critical community often have different "takes" on a problem and the emphasis that should be given to its various causes, underscoring the multiplicity of voices that construct an alternative perspective. He also calls attention to the fundamentally challenging nature of a critical community's ideas. For example, in explaining what differentiates a critical community from an "epistemic community"—a concept developed by political scientists to denote a group of experts who have a shared worldview—Rochon writes: "Critical communities are *critical*. They develop alternative challenging ways of looking at an issue, and their perspectives are critical of the policy establishment rather than being oriented toward helping it to function better."[8] It is this that makes them into central agents of cultural change.

In building on Rochon's notion of critical communities, we analyze the *thinking work* of participants in the anti-biotechnology movement to reveal both how these individuals came by their ideas and how they consolidated these ideas into a critical social analysis through an interactive and collective process. We illustrate how people collectively created a grievance around agricultural biotechnology through a process we refer to as "thinking as social action." We also show that there was a reflexive relationship between the process of idea generation and the thickening and expansion of social networks that provided the organizational core of the movement. As these individuals worked together and developed and spread their ideas, they attracted younger colleagues and protégés into the movement.

While useful for understanding how an oppositional analysis develops in a society, Thomas Rochon's notion of a critical community lacks a sense of the power and role of normative commitments in leading people to develop alternative cultural perspectives. Thus, another way in which we build upon Rochon's work is to show that critical communities are also *normative communities,* in that they are powerfully motivated by ethical and moral concerns. This moral dimension to the process of idea generation became a catalyst of the anti-biotech movement for two reasons. First, the moral outrage that many activists felt about genetic engineering created a powerful bond among them, even when the source of that outrage varied from corporate greed to technological overreach and the ethics of "playing god" with nature. This outrage generated cohesion among a diverse set of actors. Second, their ethical concerns underpinned a profound commitment to the issue, creating a strong feeling among activists that one had to do *something* about these new technologies, no matter what the odds were for change or how long it would take to achieve it. Such a profound level of commitment helps to explain why the activists at the heart of our story stuck with the biotechnology issue for upward of twenty-five years.

The Origins of Resistance

As sociologists Ron Eyerman and Andrew Jamison note, the content of a social movement's consciousness, or its "cognitive praxis," is always deeply rooted in particular historical and political contexts.[9] As we argued in chapter 1, the context out of which the anti-biotechnology movement arose was the 1960s and the plethora of social movements and causes that emerged during and after that politically tumultuous decade. These movements stringently opposed many of the political, economic, and technoscientific developments occurring in society and created a climate that fostered a new kind of thinking about North–South inequalities, capitalism as a socioeconomic system, the "military–industrial complex," and the environment. Organizations such as War on Want and Oxfam in the United Kingdom, the Institute for Food and Development Policy in San Francisco, and the International Organization of Consumers Unions in Penang, Malaysia, developed a powerful criticism of northern nations' development policies and the effects they were having on countries in the global South. The environmental movement raised questions about the ecological crises associated with industrial production and mass consumption. The anti–Vietnam War movement, as well as the European peace and anti-nuclear movements, not only offered an unparalleled mass challenge to U.S. militarist foreign policy but also brought close attention to bear on the technologies of war. These historical conditions interacted with the subjective worlds of individuals to generate a group of people who were predisposed to think critically about the new gene technologies even before they came on the scene in the 1970s. In other words, by the time scientists discovered how to cut and move genes from one organism to another, a set of people was already poised to look at those technologies with skepticism, suspicion, and concern.

Early developments in and around the field of genetic engineering provided grist for these concerns and helped to shape the sensibilities of a nascent group of critics. One of these developments was the rapid commercialization of biotechnology following Cohen and Boyer's gene-splicing breakthrough. Between 1979 and 1983, more than 250 small biotechnology firms were founded in the United States, and a dozen or so multinational corporations began investing heavily in these new technologies, as we noted in chapter 2. Another was the changing legal and regulatory framework governing the ownership of these interventions in nature. In one especially significant shift, the U.S. Supreme Court ruled, in the 1980 case of *Diamond v. Chakrabarty,* that genetically engineered microorganisms are legally patentable. As observers quickly realized, this meant that life itself could now be subject to exclusive monopoly patents, so long as the intervention met the standard criteria of patentability: novelty, utility, and nonobviousness. These trends caught the attention of citizens and activists who saw technology as inescapably bound up with social relations and who felt deeply uneasy about what private enterprise and the state could do with the technology. Unlike industry, which hailed the *Chakrabarty* decision as a crucial step forward, critics saw this legal development as an "enclosure of the commons" and an extension of the capitalist commodification process into a qualitatively new realm.[10]

Of course, the historical moment alone never fully explains the emergence of people who think in ways that challenge the status quo; "personal biography," or the experiences people have during their lifetimes and how they interpret them, is also important. James Jasper has perhaps gone the furthest in specifying how and why some people are stirred to take action and in relating this decision to personal experience. As Jasper suggests:

> Our cognitive beliefs, emotional responses, and moral evaluations
> of the world—the three subcomponents of culture—are inseparable
> and together these motivate, rationalize, and channel political action.
> Beliefs and feelings emerge from many sources: professional training
> as an engineer or an economist; hobbies such as gardening or medieval
> history; reading to one's child or caring for one's elderly parents;
> interpersonal dynamics that were thwarted or nurtured in childhood.
> Because everyone has a unique biography, different elements of the
> surrounding culture come to be embodied in the subjective worlds of
> individuals (through processes I'll label biographical). The ensemble
> of one's activities (past and present) makes certain feelings salient,
> certain beliefs plausible, certain moral principles more important than
> others.[11]

Coming to embrace a critical view of society—and ultimately, of genetic engineering—occurs in different ways for different people. For instance, some are exposed to a critical perspective by virtue of growing up in families where discussions of labor union struggles, a history of persecution, and the scourge of fascism are common dinner-table fare. One prominent anti-biotechnology activist recalled growing up in New York City with a mother who spent her time acting in plays about life in the concentration camps and the experience of poor immigrants in America. Another told the story of how when he was growing up, his father used to read him stories of the prophets. The message he carried into adulthood from those stories was that "when the king does something wrong, you're supposed to stand up to the king, and say, 'You're doing something wrong, and you better stop that.'"[12]

People's lived experiences are also important shapers of political sensibilities. One interviewee described her childhood in a Spanish-speaking Texas border town where class and ethnic differences were starkly apparent. A high school year abroad in Colombia had a transformative effect on her thinking.[13] Others were profoundly affected by their participation in student movements, anti-dictatorship struggles, and social justice movements. Nicanor Perlas, a Filipino who became active in the anti-biotechnology movement in the 1980s, noted, "[My worldview] changed drastically when I realized that [my] sheltered and privileged life was totally empty and meaningless amidst the sea of poor and oppressed people that was and is the Philippines." After he turned eighteen, Perlas devoted himself to social issues, including the struggle against agricultural biotechnology.[14]

Early Activism and Early Concerns

The earliest activism around genetic engineering arose out of two very different sorts of concerns and two very different communities. The first set of concerns included worries about the dangers of this novel technology to human beings and other living things and the social, moral, and ethical issues raised by intervening in nature with such a powerful new set of tools. Critically minded scientists, environmentalists, and technology skeptics primarily voiced these concerns. The second cluster of concerns involved what came to be known as the "seeds issue," or the loss of genetic diversity in the "gene rich" global South and the growing corporate control over seeds, one of the basic means of meeting human needs. The people who identified and organized around this issue came from the community of development critics. Although these two groups came to focus on bio-technology through distinctly different routes, they shared many concerns and organized together on a number of different issues.

Critics of the Technology

Perhaps not surprisingly, some of the first people to question the wisdom of genetic engineering were those who were the closest to it, namely, scientists based in the biological sciences. In fact, shortly after rDNA was discovered, a Cold Spring Harbor virologist by the name of Robert Pollack raised the specter of novel pathogens arising out of the use of the new gene-splicing technologies, and several very prominent scientists published a letter of warning about the potential hazards in *Science* magazine in 1974.[15] A year later, these scientists convened a major scientific meeting on the safety of these new technologies at the Asilomar Conference Center, in Pacific Grove, California, to discuss these concerns and to outline a set of guidelines for the conduct of recombinant DNA research.

While most of the scientists who participated in the "rDNA debates" soon abandoned their worries about gene splicing, a few remained deeply concerned about the technology's safety as well as its potential societal impacts. Among them were Dr. George Wald and Dr. Ruth Hubbard, both professors of biology at Harvard University; Dr. Erwin Chargaff, professor emeritus of biochemistry at Columbia University; Dr. Jonathan King, professor of biology at the Massachusetts Institute of Technology; Dr. Liebe Cavalieri, professor of biochemistry at the Graduate School of Medicine at Cornell University and member of the Sloan-Kettering Institute for Cancer Research Center; and Dr. Stuart Newman, an assistant professor of biology at State University of New York–Albany. Some had been associated with the group Science for the People (SftP), an organization started in the late 1960s by scientists and other academics who were concerned about the moral and social responsibilities scientists had to society.[16] Many SftP members had been active against the Vietnam War and were strongly opposed to their universities' complicity in the war effort. They also had a critical perspective on the role of capitalism in science and society.[17]

Thus, when these scientists saw their universities rushing to build rDNA laboratory facilities and their colleagues getting involved in new business ventures based on genetic engineering research, they questioned the forces driving these developments. They also asked whether enough was known about the risks and dangers of these technologies to plunge so blithely ahead.[18] As biologist Liebe Cavalieri noted in a *New York Times Magazine* article in 1976,

> Most problems of modern technology build up visibly and gradually and can be stopped before a critical stage is reached. Not so with genetically altered bacteria; a single unrecognized accident could contaminate the entire earth with an ineradicable and dangerous agent that might not reveal its presence until its deadly work was done.[19]

Another member of this group, Dr. Stuart Newman, expressed his concerns in these terms:

> In the late '70s, there began to be this ferment around recombinant DNA research. . . . Though starting out as a physical scientist, I had done a couple of postdocs in biology. . . . And from my "systems viewpoint" in biology, it seemed to me that making genetic modifications in organisms was not trivial; it had the potential of disturbing systemwide properties, even small, genetic changes. I was concerned about the effect . . . of releasing genetically modified microbes into the environment. I was [also] concerned about the influence on human health of generating new kinds of microbes.[20]

MIT biologist Jonathan King spoke even more bluntly about the dangerous role science could play in society if it was not subjected to critical examination. Doing science for the sake of science could generate enormous risks, in his view, risks that were often downplayed by those who had personal or political interests in that type of scientific research. "I was a graduate student at Cal Tech during the [Vietnam] war years, where there were a lot of missile engineers," noted King, at a National Academy of Sciences Forum on research involving recombinant DNA.

> A number of us were concerned that these people were using their scientific skills to design devices to kill people. And we would raise questions sitting around the dormitory, and they would say you are interfering with our freedom of inquiry. What freedom of inquiry? You are making missiles. They would say we are not making missiles; we are studying the motion of an elongated projectile through a liquid medium, and if we cannot do that we cannot learn about it.
>
> We are being told that if we don't want to have this experiment done, modifying living organisms to have it proved to us that it is not a disaster, we are holding back knowledge. Now I ask you, what is going to happen if by some small chance . . . the Walds, and the Hubbards, and the Chargaffs, and the Cavalieris are right; the experiment is done, and we get the answer—a disaster. Where we will be?[21]

In 1978, Francine Simring, who was a homemaker and environmental activist with Friends of the Earth–New York and who had also become very interested in the issue, urged several of these scientists (including Hubbard, King, and Newman) as well as others to join her in founding a group called the Coalition for Responsible Genetics.[22] After a few years, the coalition renamed itself the Committee for Responsible Genetics (CRG), an organization whose stated goal was to "discuss, evaluate, and educate the public about the social implications of biotechnology."[23] By this time, the group had expanded to include several more academics, a labor movement leader who was concerned about the use of the new genetics and workers' rights, and a number of environmental and community activists. For the next twenty-five years, CRG members sought to weigh in on policy discussions, influence legislation, educate the public about genetic engineering and initiate public debate on it, and offer a steady stream of analysis of a broad range of biotechnology-related issues in its bimonthly newsletter, *GeneWatch*.[24]

The members of the Boston-based CRG were not the only ones who began to worry about the meaning of the new gene technologies in the 1970s. Two other figures who became centrally involved were Jeremy Rifkin and Ted Howard. Like many of the CRG's members, Rifkin and Howard had been radicalized during the Vietnam War years. Rifkin had grown up in a politically Democratic but socially conservative working-class community on the South Side of Chicago. He was attending the University of Pennsylvania and was actively involved in university government and fraternity life when the war started, setting him on a new path.[25] After Rifkin and Howard started writing for a small, left-leaning magazine, they learned that some pharmaceutical companies were working with rDNA technologies. They did some investigative journalism and wrote a book called *Who Should Play God? The Artificial Creation of Life and What It Means for the Future of the Human Race*, which quickly became a best seller.[26] Around this same time, Rifkin and Howard established the Foundation on Economic Trends (FoET), which became a key site for criticism of biotechnology.

While some found *Who Should Play God?* to be exaggerated and alarmist, it contained an influential political–economic and philosophical critique of the new trends in the biological sciences. From Rifkin's perspective, "it was clear from the get-go that this was [going to be] the next major philosophical, scientific, technological, social, and cultural revolution."[27] Moreover, that revolution "demanded to be thought about before, rather than after, it transformed the world." Rifkin and Howard thus undertook an analysis of that revolution in all of its imaginable manifestations. Through his impassioned writings, electrifying speeches, and legislative challenges, Rifkin's voice became one of the most widely heard on the issue.

As news of rDNA spread through the global scientific community, a number of scientists, social scientists, and food and environmental activists in western Europe also began to look with a critical eye at the social, environmental, and animal and human health issues raised by genetic engineering. As we detail in chapter 4, Germany was one of the first countries in which a core group of critics emerged, sensitized by their country's use of eugenics during World War II. In the mid-1980s, a group of German feminists criticized the scientific reductionism inherent in the new genetics and voiced their concerns about reproductive rights and embryo research. Members of the German Green Party questioned the use of genetic engineering on a variety of grounds, ranging from ethics to scientific uncertainty and risks. One particularly active member of that group was a long-time activist and member of the European Parliament for the German Green Party, Benedikt ("Benny") Haerlin. Haerlin helped found the Gen-ethisches Netzwerk (Gen-ethical Network) in Berlin, an organization that served as an information clearinghouse on gene technology and reproductive medicine and helped those with a critical bent to connect with one another.[28] He also played a crucial role in ensuring that critics of the technology would have their voices heard by other members of the European Parliament, as we shall see.

In the United Kingdom, a small group of food activists, animal rights activists, critically minded scientists, and "science and technology" scholars all converged on biotechnology in the 1980s as well. Among them were Tim Lang and Eric Brunner, a food activist and a biochemist who worked with a group called the London Food Commission; Joyce D'Silva, an animal rights activist who worked for Compassion in World Farming; Sue Mayer, a veterinarian who became Greenpeace's scientific adviser on biotechnology; and David King, a geneticist who directed a small Green Party–inspired organization named Genetics Forum. Together, these and other European activists organized a multifaceted campaign to stop the biotechnology industry from introducing into Europe bovine somatotropin, a genetically engineered hormone developed to stimulate milk production in cows.

The "Development Critics" and the Seeds Issue

Several others began to question the meaning, implications, and commercial developments associated with the new biotechnologies from a totally different vantage point. In the late 1970s, Cary Fowler, Hope Shand, and

Pat Mooney began working together on the loss of genetic diversity associated with the spread of industrial-style monoculture. All three had been part of a growing community of international development critics when they stumbled upon the issue. Within several years of meeting one another, they founded and staffed a small organization called the Rural Advancement Fund International, with offices based in Saskatchewan, Canada, and Pittsboro, North Carolina.

Fowler first learned about the problem of genetic erosion while writing the book *Food First: Beyond the Myth of Scarcity* with Joseph Collins and Frances Moore Lappé.[29] *Food First* was one of the first major books to criticize mainstream approaches to development. Fowler became so alarmed that he continued to research the phenomenon when he joined the staff of the Rural Advancement Fund International in their North Carolina office, where he wrote an essay that offered the first political–economic analysis of the problem.[30] In 1976, Fowler was invited to give a talk in an undergraduate seminar at nearby Duke University. Sitting in the room was Hope Shand, a college senior who had lived in Latin America for a time and was writing her thesis on multinational corporations and the problem of hunger there. As Fowler presented his critical take on development, the Green Revolution, and world hunger, Shand became engrossed. "It was amazing," she recalled. "I just could not believe what he was saying. I could barely even find books that related to the subject I was looking at. . . . I felt like, this guy knows exactly about all the things that I am trying to write about." A year or so later, she became a VISTA volunteer and went to work with him, beginning a collaboration that would last for many years.[31]

During this same period, Fowler's path crossed that of Pat Mooney, a Canadian activist who had been working on development issues for years. Mooney first heard about genetic erosion while backpacking around the world with his wife in the mid-1970s. During his travels, OxFAM–UK asked him to investigate reports of severe malnutrition on Sri Lanka's recently nationalized tea estates. Because of the sensitivity of the topic, he had to be smuggled onto the tea plantations at night by neighboring rice farmers to talk to people. "The rice farmers told me about the problem of genetic erosion in the rice crop, saying that they couldn't get their old traditional varieties back. They didn't like . . . the Green Revolution varieties and were not happy with the situation they were in," he recalled. Although he did not think much about it at the time, he remembered the experience when he heard Fowler tie the loss of genetic diversity to the adoption of Green Revolution varieties in the global South. Convinced of the seriousness of the genetic erosion problem, he started to devote himself full-time to working on it.[32]

Together, Mooney, Fowler, and Shand formed a powerful trio working on the seeds issue. As Fowler described it,

> Modern agriculture needs predictability; therefore, plant breeders strive for uniformity. Plants are bred and inbred to develop the desired characteristics. The result has been the creation of new varieties that are extremely genetically limited. . . . Where thousands of varieties of wheat once grew, only a few can now be seen. *When these traditional plant varieties are lost, their genetic material is lost forever. Herein lies the danger.* (emphasis added)[33]

This narrowing of genetic diversity would only be exacerbated as countries adopted legislation that extended intellectual property rights over new plant varieties, argued RAFI activists.[34] In their view, intellectual property rights, especially the incipient trend toward plant patenting, posed a serious threat to the food security of the poor, because it threatened to choke off access to seeds from those who could not afford to pay. What is more, the proposed IPR legislation reflected a tremendous inequity: Although the vast amount of genetic diversity came from the global South, the effort to privatize the fruits of agricultural research would clearly benefit researchers and seed companies in the global North, who had the institutional capacity and incentives to utilize these legal protections. RAFI also predicted—correctly, it would turn out—that the expansion of IPR in agricultural research would greatly increase agribusiness's interests in the seed industry.[35]

Building on his relationships with European development groups, Pat Mooney established a "seeds campaign" in the early 1980s to spread the word about the loss of biodiversity and the issue of intellectual property rights. Mooney organized the campaign under the auspices of the International Coalition for Development Action (ICDA), an umbrella organization of development-oriented NGOs based in Amsterdam. Its earliest political efforts focused on trying to stop plant breeders' rights legislation (a "soft" form of IPR) from being adopted by national governments around the world and on pressuring the UN's Food and Agriculture Organization to improve the public gene bank system so that existing germplasm would be better conserved.[36] Later on, Mooney's successor at the ICDA, Henk Hobbelink, moved the campaign into a new organization named Genetic Resources Action International (GRAIN), based in Barcelona, Spain. Together with his French colleague, René Vellvé,

Hobbelink and GRAIN continued to work on the seeds issue for the next two decades.

The specific concerns that RAFI and GRAIN activists raised as they sought to communicate the seriousness of the loss of traditional crop varieties, the privatization of a once freely available resource (seeds), and efforts by multinational corporations to buy up seed companies around the world resonated strongly with a number of people in the global South who worked on agriculture, rural development, and social justice issues. When Daniel Querol (a Peruvian agronomist who worked on genetic conservation in Nicaragua), Camila Montecinos (a Chilean agronomist who worked with indigenous communities), and Vandana and Mira Shiva (two Indian critics of industrial agriculture) interacted with critics of the seeds issue, they found common cause and became part of a small but growing global anti-genetic-engineering network.[37] At the same time, they helped draw northern activists' attention to the importance of indigenous knowledge systems and the part that small-scale farmers had played in generating and maintaining genetic diversity in agriculture. It was not only that the third world was the source of most of the world's genetic diversity, these agricultural specialists argued, but also that this diversity was the product of millennia's worth of farmers' effort, as generation after generation of them had carefully selected, saved, handed down, and exchanged seeds with other communities. Failure to recognize the role these farmers played in maintaining biodiversity only added insult to the injury of IPR legislation, which effectively legalized the practice of "biopiracy." Indeed, if one issue offended activists' sensibilities across the board in the global South, it was the trend toward plant patenting.

For these and other southern activists who came onboard in the 1980s—people such as Nicanor Perlas and Martin Khor from the Philippines and Anwar Fazal and Martin Abraham from Malaysia—it was a short step from their existing criticism of agriculture's last "miracle" technology, the Green Revolution, to criticism of its current one, the gene revolution. Virtually as soon as these individuals heard some scientists and the biotechnology industry proclaiming that genetic engineering was going to solve the problem of world hunger, they rejected these claims out of hand. Not only had world hunger *not* been ended when governments, agricultural scientists, and extension agents introduced the Green Revolution's high-yielding varieties of wheat and rice into farmers' fields, but also these same varieties had been associated with many deleterious social and economic consequences for the poor. Farmers who could not afford the expensive inputs that the Green Revolution varieties required were impoverished by the loans they took out to try to pay for them, which resulted in many people being driven off the land after these varieties were introduced and a displacement of rural people to the cities. Highly cognizant of these phenomena, these activists criticized the whole model of industrial agriculture and the way of thinking that undergirded it. In one writing on the topic, Perlas noted,

> Agricultural biotechnologies presuppose and build upon . . . the current ideas, values, social structure, techniques and practices of capital and chemical-intensive agriculture. . . . The implicit goals and values embedded in modern agriculture subtly govern every step in the development of the new agricultural biotechnologies. Unless there is a drastic revision of much of the current thinking and practices . . . one can . . . expect to find an aggravation and acceleration of the current environmental and social problems associated with modern agriculture.[38]

Perlas was not alone in rejecting the values and beliefs underlying Western "technoscience" and in urging his compatriots to think more holistically about the relationship between human beings and the environment. Many others felt the same way.

Developing an Analytical Framework

As these individuals' work, social concerns, and worldviews intersected, they began to form a coherent analysis of the technology. The core activists who laid the intellectual foundation for the anti-biotechnology movement spent an enormous amount of time engaged in what Pamela Oliver and Hank Johnston refer to as "thinking work."[39] They sought information from a wide spectrum of sources, they read voraciously about the science and business of genetic engineering, and they critically analyzed the material they amassed. In short, they did what scholarly researchers do: they took their analytical skills, theoretically informed ideas, and knowledge of a field and applied them to a problem and a body of data. The primary difference between their work and that of scholarly researchers was that while the majority of research conducted in academia, think tanks, and research establishments tends to reproduce hegemonic kinds of knowledge and power relations, these movement thinkers stood dominant interpretations on their head. In the process, they performed one of the most important roles that social movements play in society: they thought new thoughts, generated new ideas, and created new knowledge.[40]

:ctual effort these activists devoted to studying biotechnology
:vident in their writings. Pat Mooney brought together a
)f information to make an original and powerful argument
e Earth, the first book to explore the implications of genetic engineering on biological diversity and small-scale farmers. Peter
Wheale and Ruth McNally, the British coauthors of *Genetic Engineering:
Catastrophe or Utopia?* described how they studied technical bulletins
and other obscure documents as they sought to make sense of the scientific and economic developments associated with the new technology.[41] Jack Doyle's popularly written book *Altered Harvest: Agriculture,
Genetics, and the Fate of the World's Food Supply* provides a detailed
analysis of changes in intellectual property law, the reorganization of
the pharmaceutical, petrochemical, and agrochemical industries into a
life science industry, and the efforts that life science firms were making to acquire seed companies. Unlike those who saw fantastic economic
and scientific potential in these trends, Doyle pointed to their downsides:
greater corporate control of agriculture, a loss of biodiversity, and further
erosion of farmer autonomy.

As they analyzed biotechnology, these early critics made connections
between the past and the present and between the present and the future.
As already noted, Green Revolution critics linked the socioeconomic and
environmental problems generated by that earlier agricultural technology to the probable effects of the gene revolution. Other activists drew
on critical analyses of agrochemicals and nuclear power technologies
to generate a critique of the health and environmental risks associated
with recombinant DNA. Terri Goldberg, the first executive director of
the Council for Responsible Genetics, described this process of making
connections:

> This was way back in the '70s, when recombinant DNA research was
> emerging and there was a lot of concern about . . . how this research was
> evolving, where it was going, what implications it would have, [whether
> we] were prepared to handle any of the unforeseen consequences of
> the research. . . . And so there were all these questions. . . . Most of the
> people I'm talking about had been schooled in the anti-war movement,
> had been opposed to the war . . . and saw this emerging technology as
> part of what was going on at that time. . . . There was a lot of activism
> about nuclear power, about chemical waste sites, and hazardous waste
> problems at Love Canal, those kinds of things. There was this strong
> feeling like, "Why didn't we question some of these developments . . .
> when we *could* have had an impact?" . . . Maybe at the outset of the

chemical revolution and the nuclear revolution, had we said, "Wait a minute. What are the health and safety issues here? What are the environmental implications of this?" . . . maybe we wouldn't be where we are now, trying to deal with Three-Mile Island and Love Canal and all these things.[42]

In the 1980s, the deregulatory thrust of the Reagan and Thatcher administrations, expressed in the new neoliberal ideology and in concrete policy changes that constrained government regulators and unbridled corporations, formed an important backdrop for activists' concerns. Critical observers worried that no one was protecting the public interest in food safety and the environment in the name of the "free market." For activists from the global South, their countries' histories of exploitation at the hands of U.S. and European states and firms served as a foundation for interpreting how these new technologies would be used, who would control them, and whom they were likely to benefit.[43] These activists were acutely aware that no technologies come into a social void but, rather, are introduced into concrete, historically shaped political and economic contexts. They had no reason to believe that, in a context of severe power inequalities, the new biotechnologies would produce an outcome substantively different from what other "first world" technologies had yielded previously. To repeat their conclusion from the Bogève meeting, "Like any other technology, [biotechnology] is inextricably linked to the society in which it is created and used, and will be as socially just or unjust as its milieu. Therefore, we conclude that in today's world this most powerful new technology is more likely to serve the interests of the rich and powerful than the needs of the poor and powerless."

The Social Nature of Activists' "Thinking Work"

In theorizing how knowledge is produced, Ron Eyerman and Andrew Jamison note that knowledge reflects "the product of a series of social encounters, within movements, between movements, and . . . between movements and their established opponents."[44] But before this final product can be achieved, activists must first engage in their own internal discussions, which help them to develop a collective critique of an issue. This is exactly what happened in the early phase of the anti-genetic-engineering movement, when people met to talk and to *think,* that is, to hash out their ideas with others who shared the same basic worldview. In the course of these encounters, each participant brought his or her own particular concerns to the table, so people exposed one another to new ideas, learned from one another, and broadened their understandings. In the process, those who constituted the activist lifeworld developed a more coherent, sophisticated, and multifaceted analysis, as well as a growing commitment to the issue.

Anti-biotech activists' participation in local, national, and transnational networks played a crucial role in the development of a collective critique of biotechnology. From the very beginning, the handful of activists, scientists, and academics who worked on the issue networked with and influenced one another. Jack Doyle, a U.S. environmental activist who was studying the energy sector in the early 1980s, learned about Royal Dutch Shell's motivation for investing in the seed industry from a conversation with Cary Fowler.[45] This led him to begin research on the biotechnology industry, an effort that culminated in his widely read book *Altered Harvest*. Joyce D'Silva, director of a British organization called Compassion in World Farming, heard about the use of recombinant bovine growth hormone (rBGH) from Michael Fox, of the U.S.-based Humane Society; she then worked with Tim Lang and Eric Brunner of the London Food Commission, several farm organizations, and others to develop a holistic analysis of bovine growth hormone and its hazards. The cofounder of GRAIN, Henk Hobbelink, started working at ICDA because he was interested in the coalition members' critical take on agriculture in developing countries, and then he discovered Jack Kloppenburg's and Martin Kenney's political–economic analyses of biotechnology, which provided him with a compelling analytical framework.[46]

Stuart Newman, one of the founding members of the Committee for Responsible Genetics, described the social nature of the idea development process:

> To me this was one of the peak experiences, [these] twice yearly meetings with this group of people, where we developed this critique. We started almost from nothing and we took every issue. Back then, there were not even any prospects of genetic engineering of crops . . . but we had anticipated it. And the genetic engineering of humans, well, [that] actually hasn't happened yet, [but] I started talking about it twenty years ago. . . .
>
> On the Friday before our board meetings . . . we would have [non-CRG] people from the Boston area . . . come to a series of working groups. . . . And so we would always have new people coming in. We had . . . a number of people that were involved in disability rights, and we got a real education on the eugenics . . . of ordinary medical genetics. And that was really an eye opener.
>
> It was just an incredible educational experience, and we [all] . . . just

fed off it, and it enhanced our ability to be educators and writers. . . . Each of us would be called to speak on this or that at various points. So there was a depth to our own analysis; people were working on issues that we weren't individually working on. We saw it as a whole . . . and not just little parts of it.[47]

Intellectual encounters of the sort Newman describes occurred within organizations, among organizations, and among individuals and groups based in different countries and on different continents. In Great Britain, activists came together under the rubric of the anti-rBGH coalition and the Genetics Forum to exchange ideas and develop organizing strategies.[48] In Penang, Malaysia, activists working on consumer issues, human health and nutrition, social justice, and environmental problems got together to talk about genetic engineering and to forge a collective analysis. The German Green Party's working group on biotechnology served a similar function (see chapter 4). Activists from almost a dozen different U.S.-based groups formed the Biotechnology Working Group (BWG) in the late 1980s, which became an important intellectual space for the movement in the United States. The face-to-face interactions among the members of these critical communities helped create the personal relationships and strong sense of commitment, solidarity, and mutual support that sustained this fledgling movement and made it hum with energy, tension, humor, and excitement. For many people, these encounters were an important source of inspiration and morale building. "I have really fond memories [of the BWG], because initially it was really a wonderful group," one member recalled. "I mean, I've been to some [other] meetings, and people go, 'Oh, this was like the BWG in the old days.'"[49]

From the beginning, ideas circulated around the world as critics of the technology sought contact with one another, such as the exchanges among member groups of transnational organizations. Transnational organizations such as the International Organization of Consumers Unions (IOCU) and the Pesticide Action Network (and later on, Greenpeace and Friends of the Earth) had offices in multiple countries, and their members regularly came together to discuss their work on various issues. International activist conferences, such as the Bogève meeting, represented another important site of intellectual exchange. Through such encounters, activists widened the scope of the movement's knowledge base and intellectual community, facilitating the development of an encompassing analytical perspective.

Idea and information exchange also occurred as people moved back and forth across continents. For instance, Benny Haerlin learned about genetic engineering from Linda Bullard, a co-worker of Jeremy Rifkin, during a 1986 visit to the United States. After returning to Brussels, he discovered that a number of German feminists and scientists were already working on the issue, and he joined them. Soon thereafter, he hired Bullard to come to Brussels to become the biotechnology adviser to the Green Party. The intellectual exchange among Rifkin, Haerlin, Bullard, and other critics of genetic engineering and life patenting continued over many years and played a role in helping to make biotechnology into one of the central political and legislative issues occupying the Greens.[50] The international seeds network that the ICDA and GRAIN built across continents was another vital arena of people, information, and idea sharing, as was the "No Patents on Life" campaign, which activists established in Europe (see chapter 4).

The Dynamism of Activists' Ideas

The various analyses and publications produced by CRG members, Rifkin and Howard, RAFI activists, and others clearly reflected a particular framing of the issue, in that they "selectively punctuate[d] and encode[d] objects, situations, events, experiences, and sequences of actions," amplified certain values and beliefs, attributed blame, and identified causes.[51] But these analyses and publications represented much more than that; they also reflected a social process of idea formation, elaboration, and reformation. These were the products of raw intellectual work, as viewed through a particular set of lenses. This point is underscored by the highly dynamic nature of the analyses these individuals generated, which emerged synchronically with advances in the science, changes in the law and regulatory system, and developments in the biotechnology industry itself. No sooner would a development occur than one or more members of this critical community would be thinking, talking, and writing about it.

In one particularly revealing example of this moment-by-moment intellectual work, Jeremy Rifkin told us, "I sat in the Supreme Court when the hearing was held on *Chakrabarty* [the landmark Supreme Court decision that allowed microorganisms to be patented]. There were only a few of us there, and I knew that that would be the commercial begetting of the next two centuries."[52] Another trend these activists identified as it was unfolding was the movement of large pharmaceutical, chemical, and petrochemical corporations into the agricultural sector and these companies' efforts to buy up seed companies. "Let us state the problem unequivocally: The greatest threat in the new biosciences is that life will become the monopoly property of a few giant companies," Fowler and his colleagues wrote in their two-hundred-page summary of the Bogève workshop, as they watched these changes unfold over time.[53]

Not only did the scope of the movement's analysis grow as activists worked on the issue; so did their capacity to analyze relevant developments. People relied on a combination of formal training, personal interest, and experience to become experts on specific aspects of biotechnology. Such was the case with Michael Hansen, from Consumers International, who developed a rich knowledge of the toxicological issues associated with GM food safety, and with the London Food Commission's Eric Brunner, who became a specialist on bovine somatotropin's biochemistry. In similar fashion, Martin Khor and Chee Yoke Ling, of the Third World Network, Beth Burrows, of the Edmunds Institute, and Kristin Dawkins, from the Institute for Agriculture and Trade Policy, all accumulated considerable knowledge of international law and environmental safety concerns related to GMOs as a result of the many years they worked for an international "biosafety protocol" to regulate trade in transgenic organisms.[54] In part, as a result, they were able to parlay this knowledge into effective political pressure at the United Nations.

Motivating Cognition: Morality and Value Commitments

Like the industry scientists and businesspeople described in chapter 2, most of the activists who began organizing around biotechnology came to the issue with a particular worldview. From within this worldview, certain ideas held by scientists, industry, governments, and universities and their accompanying behaviors were deemed acceptable and "right" by these individuals, while other ideas and behaviors were seen to be unacceptable and in some cases totally egregious. In fact, it was the normative elements of people's worldviews, which were inextricably linked to their values, that motivated them to take action. Exploring these moral sensibilities offers us a deeper insight not only into the content of these activists' worldviews but also into what kept people fighting for the cause, even in the face of tremendous adversity.[55]

Although the normative sensibilities and moral outrage of antibiotechnology activists had multiple dimensions—that is, not all were shared by all activists or emphasized by everyone to the same extent—they represented a unifying force for those in the movement. As is suggested by the song lyrics below, written by the British folk musician Tina Bridgman, many of those who looked on the new scientific–industrial developments with foreboding rejected the predominant values of contemporary capitalist society and embraced alternative value schemes and moral codes. These alternative schemes rejected mainstream society's utilitarian views of nature and consciously placed the well-being of society and the environment ahead of the profitability of corporations.

You Have No Right

First you come with a promise,
a silver spoon to feed the masses,
But looking at your history
You're not what you make out to be.

The problem you say is starvation,
and how can we feed every nation?
So you meddle with life and evolution,
And justify your profits.

Refrain: But you have no right, you have no right,
You have no right to make a claim on this life.

And the governments who are here to defend us
Seem to have too easily surrendered
To your hypocrisy, and I wonder . . .
Where money talks, you've really got them.

And it's my basic human right
That I have a choice, I can decide
What to leave, what to preserve.
It's the least I offer tomorrow's world.

And it's not much of a legacy
Nothing less than insanity
to jeopardize life for eternity
—You're mistaken to call this progress.

So I will walk to the places I know
Where the bird has song and the wild flower grows.
You see, I need to feel the hope
That the world sees sense and not destroy this.

Refrain: But you have no right, you have no right,
You have no right to make a claim on this life.

A major outrage for many derived from the perception that corporations had come to wield an unprecedented and dangerous degree of power

in the world, which they readily and quite typically abused. "If people th.
it's about food, I don't think so," observed one of our interviewees, a mai
in his late fifties. "I think it's about the domination of the means of reproduction of genes and the means of development, proteins. This is about 'the
corporations are done with zinc and trees,' and now, instead of owning the
field, instead of enclosing the field, you just own [the] soybeans."[56]

Another activist who had worked on the biotechnology issue for well
over two decades noted,

> We don't have a blanket [opposition] to genetically modified
> [organisms]. . . . Really it's a matter of control and ownership, and
> that's why we've always focused on those issues. Because, it's just so
> clear to me that . . . when . . . you look at what has really happened
> in the 130 million acres that have been planted, and the fact that
> Monsanto's technology accounts for 91 percent of [that] worldwide,
> what else do you have to know? We're talking about control by a
> single company. There are not really even five gene giants. There's
> really only one.

A third activist, a woman in her midfifties, observed, "[Biotechnology]
comes on the scene when corporations have too much power in the
world. Especially in the global environment, where there aren't in fact
any governance structures, it is especially dangerous, because it is going
to be driven with greed and opportunity."

Such sensibilities about corporate greed and "overreach" were heightened by the unsavory reputations of the companies developing this technology, stemming from their pasts as arms and chemicals producers. For
those who came of age during the 1960s and 1970s, names such as Dow
Chemical, Monsanto Corporation, and Imperial Chemical Industries
conjured up an image of firms that had developed dangerous chemicals
such as Agent Orange and Paraquat, not of companies interested in ending world hunger. The strong distrust these individuals harbored toward
such companies was expressed by activists in personal and political terms.
"I've had cancer," acknowledged one of our interviewees.

> I think that pesticide companies, like tobacco companies, have been
> lying for years about the dangers of pesticides. And the fact [is] that
> they've gotten away for years with promoting these pesticides and
> denying their risks. And now, they're admitting these risks of pesticides
> and saying, "Buy biotech." You know, these are the same companies!

Many activists were deeply distressed by and indignant about the lack of
democratic input into decisions about these literally life-altering technology

'est offense for many lay in the perception that corporations
making decisions about technology choices that carried
·ions and repercussions for the rest of society, indeed, the
out any real public participation.

ιιere you're unleashing a science and a technology that we have no
idea [what it] means to the long-term evolution of the ecology. And it
just dumbfounded me. I mean, it really did stop me on a gut level, and
I said, "This is wrong." We need moral debate. We need public debate.
We need scientific debate. We're completely eliminating a democratic
process for discussing this technology.

The devotion of all these resources to biotechnology "is ludicrous," noted
another activist, lamenting the lack of discussion about the public funding of these technologies in the United States.

Because if you look at [it], where was the informed consent? Where
was the debate about biotech being [the way to go]? I mean, we all
agree that there are big problems in agriculture. . . . But where was the
public debate that said, "Let's put all our resources into biotech as a
solution to all of these problems"?

Such sensibilities about the lack of democratic process and choice
were connected to a deep anger that the giant life science corporations
had aggressively pushed for the right to patent life and had actively been
supported in their quest by the United States and, later, European governments and courts. The notion that life could or should be patented
revolted these individuals and confirmed in the clearest of terms their
analysis of the seamy side of capitalism and the limitless nature of corporate greed. Noted one activist, "This is grandiose in my opinion, in
terms of human history. . . . This is biocommerce. This is turning the living world into [a] product. Products, we become products! We become
viewed as items of utility." An activist from India exclaimed,

The thing that offends me the most about biotechnology in agriculture
is the issue of patents. For me, it is about economics. You know,
the fact that these corporations have taken over the seeds, the fact
that 80 percent of the food that is eaten in [the United States], the
germplasm of that comes from *my* side of the world, and the audacity
of these corporations then to go and destroy our biodiversity and tell
people how to grow food!

In some activists' eyes, the commodification of life was redolent of a
profoundly alienated society, one in which human beings had reached

a terrible and unacceptable state of spiritual emptiness and disconnected-ness, from one another as well as from other forms of nature. "The tech-nological imagination reflected in genetically engineered foods is one that views limits as ontologically evil," noted the director of an anti-biotechnology group in Washington, D.C. "It views all limits as evil, and as an environmentalist, as a human being, I find that very, very disturb-ing." For such individuals, the view that human and nonhuman life was something that needed to be improved upon—that *could* be improved upon—reflected the moral impoverishment and excessive utilitarianism inherent in capitalist society. "I think this is the most powerful technol-ogy we are going to confront in this century, and it does, it will, ask us what it means to be human," claimed a woman who had been working on the issue for many years. "To look at every living thing and think of it as just a big Lego set—just a bunch of genes that you can put together in different ways so they can serve our purposes better—that's not the relationship I think we should have to nature."

This array of related concerns created a powerful feeling that one *had* to do something to alter the course of history around these new genetic engineering technologies, regardless of what the odds were for change or how long its achievement would take. Indeed, one activist suggested that *not* doing anything about the host of moral and ethical issues raised by genetic engineering would be akin to choosing not to stand up when the Nazis came for the Jews:

> I'm not saying that they [the industry and scientists] are going to be successful at this. I'm saying that even as failures, the results could be catastrophic and ultimately change everything as we know it. And that's why I'm fighting cloning of humans and [other biotechnology]. Unlike some of my environmentalist friends, I don't see the line between humans, animals, and plants. You know [the saying], "I wasn't a tomato, so I didn't speak up for the tomato; I wasn't a Jew, so I didn't speak up when they came for the Jews. And then they came for me"? Well, we *are* speaking up for the tomatoes and the fish and the chickens, and now they *are* coming for us.

Another activist, reflecting upon why she had stuck with the issue for so many years, explained, "I think it is worth many lives for people to get this one right, you know, for us not to be swallowed by a corporate-determined techno-eugenic view of our future." Virtually all of these core activists recognized that the work they had taken on was not going to be completed quickly. "We started our [work] with Monsanto in 1982," observed Jeremy Rifkin. "It takes a generation. It's going to take another generation before we clean this up."[57]

These quotes not only convey an enormous depth of conviction, but they also reveal an intimate relationship between these individuals' normative views and their intellectual analysis of biotechnology in advanced capitalism. Activists' moral convictions and value commitments provided a critical impetus for their intellectual work, helping to inspire and sustain their analytical efforts over several decades, even though many felt themselves to be working on a "lost cause" and in the midst of impermeable political–economic structures. At the same time, activists' intellectual activities fueled their moral outrage as their analyses revealed patterns, connections, and structures that were inimical to their sense of a good society. In revealing the close and mutually reinforcing nature of this relationship among ideas, values, and moral principles, our analysis highlights the role played by normative concerns in the thinking work of movement intellectuals.

Growing the Movement

From the 1970s to the early 1990s, the activists described above and dozens of others sought to spread their message and introduce the issue to a broader constituency. Indeed, one of the key achievements of the earliest anti-biotechnology activists was to bring genetic engineering within the purview of organizations working on a wide range of other issues and to enable such organizations to coalesce around the potential impacts of this engineering. For example, the prospect of "deliberate release" of the technology brought citizens concerned about the environmental and health hazards of these new engineered organisms into discussions with citizens concerned about gene patenting, corporate concentration, and loss of biodiversity and livelihoods, especially in the global South. This range of overlapping concerns and coalitional possibilities was well demonstrated by the various groups who got involved in the issue in western Europe, which included the German Green Party, Greenpeace Switzerland, the U.K. Green Alliance, the U.K. Genetics Forum, the International Coalition for Development Action, and Compassion in World Farming.

European activists were particularly successful in broadening the appeal of their arguments, because they were able to place the politically powerful concepts of societal *risk* (cast in terms of health, environment, ethics, and culture) and consumer rights at the core of their analysis of the meanings of the new gene technologies. Civil society associations that focused on risk—especially environmental organizations such as Greenpeace, Friends of the Earth, and the British Soil Association, as well as farmers' organizations such as the Confédération Paysanne, in France— had already established a strong watchdog presence in European countries. As influential members of these groups read or heard the analysis of the anti-biotechnology intelligentsia and were persuaded to get involved, the movement changed form and shifted into a new phase, expanding its popular base and its repertoire of contentious actions.

As we will show in chapter 4, a key turning point in Europe came in 1996, when Benedikt Haerlin convinced Greenpeace International to initiate a major new campaign against GMOs, which he then coordinated and spearheaded. Organizing around Monsanto's first introduction of GE seeds into European markets, activists kicked off their campaign by blockading the ports into which Monsanto was importing its seed, unlabeled and mixed in with non-GE soy seed. Activists represented this as a secretive and devious effort to contaminate European seed stocks and thereby impose on consumers food technologies whose effects were both risky and ineradicable. In an atmosphere of increased public sensitivity to food issues, heightened apprehension, and suspicion of corporate motives, many European NGOs developed new campaigns around genetic engineering and began to work in coalitions.[58] At the same time, several private foundations started funding groups to work specifically on genetic-engineering issues. For instance, the Goldsmith Trust, a large environmental foundation, started financing a campaign against genetic engineering of foods in Britain, including providing support to the Genetics Forum, a coalition of thirty-one groups involved in anti-biotech work.[59]

In the United States, a thickening and expansion of activist networks also occurred as activists working on distinct but related issues began to see a need to address the potential power of this new technology. As indicated in Figure 2, the number of groups with one or more persons working on biotechnology rose from three in the 1970s to fourteen in the 1980s and to over thirty in the 1990s. The Biotechnology Working Group rapidly doubled in size, from a dozen or so people to twice that many. Several representatives from family-farm groups joined, as did one each from the National Toxics Campaign, the Minnesota Food Association, and the United Methodist Church in Washington, D.C. The individual who had the greatest impact on expanding the anti-GE movement in the United States (and to some extent, globally) was Jeremy Rifkin. Among those who worked at Rifkin's "shop" (the Foundation on Economic Trends) in the 1980s and 1990s were Nicanor Perlas, who (as already noted) became head of the

1970s

Coalition for Responsible Genetic Research (later renamed the Council for Responsible Genetics, CRG), ca. 1977 (formed by members of Science for the People and of the Coalition for the Reproductive Rights of Workers)

Foundation on Economic Trends (FoET), 1977

National Sharecroppers Union/Rural Advancement Fund International (RAFI), 1978

1980s

Council for Responsible Genetics (CRG)

FoET

RAFI

Environmental Policy Institute, 1983

Earth First! (Strawberry Liberation Front), 1986

Concerned Citizens of Monterey, 1986

Concerned Citizens of Tulelake, 1986

National Wildlife Federation, 1988

Biotechnology Working Group, circa 1988

Consumers Union, 1989 (formerly known as Consumers International, and before that as the International Organization of Consumers Unions, IOCU)

Environmental Defense, 1988

National Farmers Union, 1989

Center for Rural Affairs, 1989

Pesticide Action Network North America (PANNA), 1989

National Toxics Campaign, 1989

1990s–2000

FoET

CRG

RAFI-USA

RAFI-ETC Group (broke off from RAFI-USA; RAFI-USA continued to work on biotechnology after the split)

Union of Concerned Scientists (UCS), 1992 (biotech work began in 1992 when two members of National Wildlife Federation moved to UCS)

Center for Science in the Public Interest

Environmental Defense

PANNA

Edmonds Institute, ca. 1993

Family Farm Defenders, 1994 (based in Madison, Wisconsin; grew out of the struggle against recombinant bovine growth hormone in the late 1980s)

Pure Food Campaign (PFC), ca. 1996 (began as part of FoET and split off in 1990s; run by Ronnie Cummins, under auspices of the Organic Consumers Association)

International Center for Technology Assessment (ICTA), 1994 (started by Andy Kimbrell from FoET)

Institute for Agriculture and Trade Policy (IATP), ca. 1994

Center for Ethics and Toxics, mid-1990s

National Family Farm Coalition, ca. 1994

Greenpeace International, 1996

Institute for Food and Development Policy, ca. 1996

Mothers for Natural Law, 1996

Center for Food Safety, 1997 (started by Andy Kimbrell and Joe Mendelson from ICTA)

Institute for Social Ecology–Biotechnology Project, 1997

U.S. Public Interest Research Group (USPIRG), ca. 1997

Greenpeace USA, 1998 (involved intermittently since 1998)

Organic Consumers Association, 1998

Washington Biotech Action Council, ca. 1999

Organic Consumers Association, 1998 (started by Ronnie Cummins, formerly of Pure Food Campaign)

Northeast, Bay, Northwest, Down South, and Grain RAGE, 1998–2000 (local and regional groups; RAGE stands for Resistance Against Genetic Engineering)

Gateway Green Alliance, St. Louis, 1998

Vermont Genetic Engineering Action Network, 1999

National Environmental Trust, 1999

Friends of the Earth, 1999

The Campaign to Label GE Foods, 1999

Clean Water Fund, Boston, 2000

Minnesota Genetic Engineering Action Network, 2000

Genetic Engineering Action Network, 1999

Sierra Club, 2000

Figure 2. Growth of the U.S. anti-biotech movement, 1975–2000: organizations (with at least one staff member) working on agricultural biotechnology. Sources were author interviews; the Internet; Krimsky 1988; Tokar 2001; and published materials from the listed organizations.

Center for Alternative Development Initiatives in the Philippines, Bullard, who played a key role in connecting the U.S. and Europ anti-biotechnology movements to one another; Andy Kimbrell and Jc Mendelson, two lawyers who moved on to form the International Center for Technology Assessment and the Center for Food Safety, in Washington, D.C.; Ronnie Cummins, who became head of the Organic Consumer's Association, based in Minnesota; Howard Lyman, who came from the National Farmer's Union and later founded the organization Voice for a Viable Future; and John Stauber, coeditor of *PR Watch,* based in Madison, Wisconsin, and coauthor of *Toxic Sludge Is Good for You.* A number of these individuals became important anti-genetic-engineering activists in their own right.

From a loosely interlocking network of activists who had come together to work on an issue of shared personal concern, this emergent social movement thus began to build a firm organizational base in public advocacy groups. The fact that many of these people were long-time activists working in established organizations or had been involved in other social movements and struggles, or both, implied two things about what Marshall Ganz refers to as a movement's "strategic capacity," or its ability to interact efficaciously with its environment and to mount effective strategy.[60] First, their years' worth of organizing experience signified that these activists had a wealth of skills and knowledge they could bring to bear in the struggle against the biotechnology industry and pro-biotechnology governments and government agencies. Second, it signified that many members of the anti-biotech movement operated from organizational bases that were stable as well as politically engaged. Indeed, many of these environmental, consumer, and other kinds of advocacy organizations had become institutionally entrenched as public watchdogs. From such strategic vantage points, anti-GE activists were well placed to do battle.

These activists also wielded a significant cache of cultural capital. Prior to the mid-1990s, when the movement experienced a major expansion, the core of the anti-GE movement mainly comprised educated, middle-class professionals, many of whom held advanced degrees in science, law, economics, or planning. Together, these personal characteristics—class, education, and experience—imbued these actors with a strong dose of confidence and a powerful sense of their own agency. This was clearly revealed in a comment made by one long-time anti-biotechnology activist who, when asked why he had become involved in such a difficult and uphill battle, exclaimed, "*People* make history, either by active involvement or by letting someone else do it for you, and maybe not in your interest. *People make history!*"[61]

g network of activist–intellectuals built on the ideas
' to develop their own specific but synergetic critiques
heir activities had two effects that were crucial for
social movement. First, they generated new publicly
;e about the technology that articulate spokespeople
lenge the claims of industry and government directly.
Second, they attracted a small but steady flow of financing from liberal
private foundations that saw research-based advocacy organizations as
playing an important watchdog role in a time of deregulation.

Lifeworlds in Conflict

Contrary to popular wisdom, the anti-biotechnology movement did not
emerge in the late 1990s, when news of colorful anti-GMO protesters in
Brussels, London, India, and elsewhere splashed across newspapers, TV
screens, and other media outlets around the world. The origins of this
movement date back to the time when the first breakthroughs in gene splicing occurred in a molecular biology laboratory in the San Francisco Bay
area. Even before that fateful moment, however, the seeds of resistance to
agricultural biotechnology were sown in a small group of academic scientists' critical reflections on the role of scientists in society; in the social
movements that emerged during the 1960s and 1970s and the environmental, health, and social justice issues that animated them; and in southern
activists' reactions to the societal impacts of the Green Revolution and the
northern aid agencies' problematic attempts to "develop" their societies.

For a certain group of people, the experiences of living through this
historical moment altered their lives and profoundly shaped their worldviews. They became part of activist communities that shared their experiences and perspectives. Like the lifeworlds of the scientists and industry
officials we analyzed in chapter 2, these activists' lifeworlds served as
a lens through which they saw, picked out, and interpreted particular
events and made connections among different phenomena over time. For
example, to those who were critical of nuclear power, it made perfect
sense that attempting to intervene in human, plant, and animal genomes
was bound to lead to unanticipated (and most likely irrevocable) problems. Similarly, no one who was critical of the Green Revolution had any
trouble seeing why the gene revolution was also likely to have problematic social, economic, and environmental consequences. In short, people's
readings of events and situations, as well as of one other, were inextricably
tied to their lifeworlds.

Both anti-genetic-engineering activists and those working in the industry developed their lifeworlds through a combination of social (and *socializing*) and personal–biographical experiences, along with their exposure to certain ideas, subcultures, and (in some cases) academic training. Furthermore, both groups' ways of looking at the world were intimately intertwined with their ethical and moral sensibilities, which classified certain behaviors, occurrences, and societal structures as right, acceptable, and legitimate and others as wrong, immoral, and intolerable. Membership in particular social networks introduced people to, and tended to reinforce, certain culturally constructed ideas and ways of thinking and rendered others "nonsensical." Thus, for example, those entering the activist community quickly became exposed to the notion (if they did not already have it) that patents on life were outrageous, immoral, and imbued with a complement of negative meanings, whereas those who were integrated into the biotechnology industry were exposed to a very different cultural reading, namely, one that viewed patents as just and fair compensation for their intellectual labor and their firms' financial investments in the research. From this cultural perspective, patents were simply part of the way things worked in their world.

One important difference between the activist community and its lifeworld, on the one hand, and the industry and scientific communities and their lifeworld, on the other, was that the former was in tension with predominant ways of thinking and in opposition to the status quo, whereas the latter reflected norms and values that were in sync with the dominant ideology and upheld the status quo. This meant that the activists' analysis of genetic engineering was not simply there for the taking but had to be constructed piece by piece, as we have shown in this chapter. This involved a tremendous amount of "thinking work" on the part of those who constituted biotechnology as a grievance in society.

CHAPTER FOUR

The Struggle over Biotechnology in Western Europe

In retrospect, it seems incredibly naive, but it's the truth. We had real leadership; we had worked hard to do it. We had shown faith in this science when others were dubious, and it all seemed to be working. So we painted a big bull's-eye on our chest, and we went over the top of the hill.

—Robert Shapiro, chief executive officer, Monsanto

Getting a new biotechnology crop approved is probably high on the dumb things to do for your stock price right now. It amounts to sticking your chin out and saying, "We're going forward with this stuff regardless."

—Alex Hittle, industry analyst with A. G. Edwards and Sons, Inc.

As the twentieth century came to an end, the juggernaut that had become the agricultural biotechnology industry came to a screeching halt. Robert Shapiro, the sustainable development guru at the helm of Monsanto, was chagrined to find that his company's strategy for moving into the global market was seriously backfiring in Europe. What in the mid-1990s looked like a moment of immense commercial promise—the U.S. and European governments had approved the commercial release of the first GM crops, the industry had moved its GM soy and corn seeds into the market, and U.S. farmers were highly enthusiastic about the new technology—by the end of the decade looked like an impending disaster. In what seemed like an overnight shift, European food processors elected to stop using GM ingredients in their food products, retailers stopped stocking their shelves with GM food, and the European Union adopted an "unofficial" moratorium on new GM crop approvals in 1999, prompted by a policy shift in several European countries. To make matters worse for biotechnology companies, many other countries around the world, particularly former European colonies that still sold the bulk of their agricultural products to Europe, carefully observed what was happening and steered clear of these new technologies. A constellation of forces had thrown a formidable barrier in the way of the biotechnology industry. .

What can explain these radical shifts in the behavior of the European food industry and in government regulatory policy? Why did the market for genetically modified food and seed in Europe suddenly dry up, throwing the industry into a tailspin and dramatically altering the trajectory of the technology? In this chapter we argue that what might look like a fairly abrupt shift in European politics and policy was actually the outcome of fifteen years' worth of continuous organizing, "counter-discursive" work, and adept political maneuvering on the part of anti-biotechnology activists as they sought to stop genetic engineering from being introduced, accepted, and adopted in Europe. In the first part of this period, that is, up through 1995, most of the movement's activities took the form of developing an alternative discourse on biotechnology (based on the collective intellectual processes described in chapter 3), challenging "expert knowledge," and trying to influence the construction of regulatory frameworks at the level of national governments and the European Union as a whole. In the process, anti-biotech activists brought their worldview into the political sphere both at the policy level and in the court of public opinion.

During the 1980s and early 1990s, most European governments, as well as the European Union overall, attempted to establish a research and investment climate that would facilitate the development of the biotechnology industry and the use of these new technologies in the agricultural, pharmaceutical, and medical spheres. As these governmental bodies sought to achieve these goals, anti-biotech activists tried to thwart their efforts in a variety of ways. One involved trying to influence the way the "deliberate releases" of genetically engineered organisms were regulated, at both the national and the EU levels. Another involved waging a protracted battle against the proposed European Commission "Patent Directive," known as the "Life Patents Directive," which would extend intellectual property rights to living organisms, much the way the U.S. Supreme Court decision *Diamond v. Chakrabarty* did in the United States. Activists also struggled to push national and supranational regulatory authorities to exercise more stringent control over this emerging technology.

During the first ten years of the struggle, the impact of the activists' actions was, for the most part, not immediately apparent. In Germany, for instance, where the anti-biotechnology forces were the most active, the German government continued to be highly supportive of biotechnology, despite strenuous efforts by the German Green Party and many NGOs to alter the government's policy stance. In Britain, the government showed its unwavering support for biotechnology by agreeing to serve as the *"rapporteur"* (or sponsor) for several of the industry's initial applications to the EU for regulatory approval, despite the activities of a growing group of activists there. The same was true of the French government, which actively encouraged the development and testing of genetically modified crops throughout most of the 1990s.

But with hindsight, anti-biotechnology activists took actions in each of these countries (as well as in Denmark, Austria, and other countries) that helped to lay a strong foundation for future success in opposing the technology. For example, during this early period, activists created the alternative discourses that helped to mold the public's thinking about biotechnology. They also initiated campaigns that forced politicians and policymakers to acknowledge that much was still unknown about the risks of these technologies. Finally, they contributed to forging a bio-technology policy at the European Union level that opened up important new possibilities for slowing down the technology's introduction. That policy established a regulatory system that was based on the *process* of genetic engineering rather than on the *products* the new molecular bio-technology techniques produced. Once a process-based system was in place, activists took advantage of its particulars and the political struc-ture of the EU to help slow down and eventually halt new crop approvals by the European Commission. Put in the language of social movement theory, anti-biotechnology activists both created and then utilized a new set of political opportunities to effect change.

After 1995, conditions for organizing changed substantially when the first GM crops started arriving in Europe and activists were able to tar-get the technology through a concrete set of commodities, namely, ge-netically modified food. Through consumer campaigns and direct action, they openly challenged biotechnology at the level of public opinion and exploited key vulnerabilities of the industry. Several factors facilitated the efficacy of the movement at this point. One was the European public's changing sentiments toward GMOs, which reflected the power of the movement's discourse on GM food as risky, unhealthy, and linked to the ills of industrial agriculture. Another was the structure of the commodity chain for GM food, which provided important weaknesses the activists could exploit. A third facilitating factor derived from the corporate culture of the leading agricultural biotechnology company, Monsanto, which combined an attitude of supreme self-confidence about the strength of its science with a highly U.S.-centric perspective about public sensibilities and the right of firms to develop and market new products as they wished. This corporate culture and worldview made Monsanto a perfect target for activists. They also rendered the company incapable of anticipating the reaction of populaces and governments with different cultural sensibilities, different senses of citizens' and consumers' rights, and different views of biotechnology from those of its own.[1] This failure of anticipation led the company to make a number of fatal mistakes in market introduction.

As is evident from this synopsis, the story we tell about what happened with GMOs in Europe locates anti-biotechnology activists at the heart of the action. Although these activists (and their NGOs) were by no means the only actors who produced Europe's policy turnaround and market closure in the late 1990s, they were the central protagonists. For it was they who led the charge against agricultural biotechnology and they who made it into a public issue. Stated somewhat differently, had the anti-GMO movement *not* mobilized around agricultural biotechnology, the technology would have enjoyed a very different reception and followed a distinctly different trajectory from those it did, both in Europe and worldwide. Not only would Europeans be consuming a much greater quantity of GMOs in their diets and have a lot more of them growing in their countryside, but the world at large would not have experienced such immense turmoil around GM food, feed, and seed. As it turned out, the planting and consumption of genetically modified crops in Europe were severely curtailed by the events of the late 1990s, and all foods containing more than a modicum of GM ingredients are clearly labeled. Europe's chief agricultural trading partner, the United States, lost important European markets, and many countries in the global South became more cautious about introducing GM seeds. While few anti-biotech activists were consciously aiming for these outcomes (most would have preferred to see the technology completely abandoned, for example), farmer, food industry, and government behavior toward agricultural biotechnology all bear the profound mark of their strategic actions.

The Development of Biotechnology in Europe

As we saw in chapter 2, the 1970s and 1980s were a time of tremendous excitement about the developments in the new biosciences, particularly molecular biology. During this period, private industry attracted molecular biologists, plant geneticists, biochemists, and others from major research universities to work on basic and applied research. The more entrepreneurially oriented of these scientists established their own biotechnology start-ups, and dozens of multinational corporations built up their own biotechnology capabilities and labs. In the process, a new biotechnology industry was born.

While the majority of the firms composing this new industry were based in the United States, Europe was also the home to a growing number of biotechnology enterprises. European chemical and pharmaceutical giants, such as Hoechst AG, BASF, Sandoz, ICI, and Rhône-Poulenc, all invested in biotechnology, excited by its economic prospects and motivated by their competitors across the Atlantic doing the same.[2] As in the United States, much of this investment was driven by an interest in research that was oriented toward pharmaceutical and medical applications, but some of it was focused on agriculture. The small, venture capital–based biotechnology firms that had sprouted up like mushrooms in the United States were significantly fewer in number in Europe but were by no means absent from the biotechnology scene.

From most policymakers' perspectives, the advances in molecular biology that had so energized the scientific community and a growing number of corporations and investors could not have come at a better time. In the early 1970s, European countries found themselves in the midst of a difficult economic crisis, thanks to oil shortages, rising inflation rates, and competition from abroad. U.S. manufacturing industries had already surged ahead of Europe's, and now the same phenomenon was occurring with the new growth industries of semiconductors, computers, telecommunications, and consumer electronics. Adding to Europe's woes was Japan's emergence as a major powerhouse in all of these sectors. To many European governments, biotechnology represented a vital new industry on which to set their sights as they attempted to reactivate their faltering economies. Unlike traditional manufacturing and other "sunset" industries of the past, biotechnology represented the wave of the future—an information technology that was poised to transform not only the economy but also numerous other aspects of life. Furthermore, its potential for helping to solve serious problems, from disease to world hunger and environmental degradation, was vast.

On the basis of this assessment, most European governments actively sought to encourage the growth of the biotechnology industry. Although biotechnology had originally been greeted with deep suspicion and considerable wariness in Europe, this perspective began to change over time, at least among policymakers. The official discourse of risk that had dominated government discussions of the issue in the 1970s gave way to a much more positive official discourse in the 1980s. In the new narrative, biotechnology was painted as a strategic, new "high-tech" growth industry, one that was both capable of and necessary for pulling European economies out of the doldrums.

In consistency with this new framing, policymakers in many European countries, as well as in the EU more broadly, threw their support behind biotechnology. Germany, for example, increased federal expenditures on biotechnology from 44 million marks in the 1970s to 92 million in 1980 and then to 256 million in 1990, with individual states *(Länder)* devoting millions more.[3] The British government, despite its embrace of a neoliberal economic ideology under Margaret Thatcher, also offered the sector funding and other sorts of support, spending hundreds of millions of pounds of public funds on biotechnology R&D in the 1980s.[4] France indicated its enthusiasm not only by backing domestic research on biotechnology but also by agreeing in many cases to serve as the *rapporteur* for biotechnology companies that wanted to have their applications for specific GMOs introduced into the EU, sponsorship being a legal requirement for approval.[5] Indeed, of the first fifteen commercial applications made to the EU for GMO approval, nine were submitted through the French government, reflecting the widely held (and initially correct) view within the biotechnology industry that the French government was highly sympathetic to its goals and supportive of its development.[6] (Britain was another country that agreed to play this role, at least for a time.) Finally, the European Commission developed a plethora of initiatives aimed at growing the biotechnology sector. These included providing funding for specific research programs, providing institutional and infrastructural support to universities, research institutes, and the private sector, and developing supportive policy proposals, such as the European Patent Directive, which, as previously stated, focused on intellectual property rights.

Despite the palpable excitement among these and other European governments about biotechnology, not every country was enthusiastic. Denmark, for example, indicated its concerns about biotechnology from very early on, as reflected in the 1986 passage of a restrictive law on genetic engineering.[7] Austria, too, had significant reservations about biotechnology. But by and large, most countries, particularly the larger and more technologically advanced ones, indicated a desire to support the development of biotechnology.

Early Opposition to Biotechnology: Domestic Struggles

Opposition to biotechnology arose in Europe in reaction to its dominant cultural framing as an unqualified "good" for society and the economy, and to the associated government and industry efforts to stimulate its development.[8] As noted in chapter 3, the seeds of struggle over biotechnology were initially sown by a handful of scientists, academics, and social activists who were part of the 1960s and 1970s generation. These were people who had participated in the wave of political mobilization that swept through most Western democracies during that period and who were as concerned about what genetic engineering *symbolized* in terms of the current and future social order as they were about its potential impacts on public health, the environment, and agriculture. This was especially true in western Europe, where genetic engineering was seen by those in the environmental, anti-nuclear, and other "quality-of-life"–type movements as a manifestation of the ills of advanced capitalist society and of technoscience-based modernization. Indeed, whereas most European governments and the biotechnology industry read and framed the advances in molecular genetics and biology as powerful engines of social and economic progress, these social movement critics read them as a deeply disturbing capitalist assault on nature, society, and the food system. All of these ideas became tied up with the notion of genetic engineering as a profoundly problematic technology.

Creating Dissensus in Germany

Within Europe, opposition to biotechnology emerged first, and most forcefully, in Germany, led by a group of middle-class German feminists, critically minded scientists, and environmentalists concerned about the normative implications of biotechnology and the nature of the risks it involved.[9] Many of the individuals who began to criticize biotechnology publicly were associated with the newly established Green Party, which formed a central node in the German (and, indeed, the larger European) anti-biotechnology movement from very early on. Largely through the vehicle of the Green Party, which was elected to the German parliament, the Bundestag, in 1983, these activists brought an oppositional perspective into the policy arena and created the beginning of a public debate around a technology that until then had been eagerly embraced by much of the country's scientific establishment, by industry, and by the government.

As Herbert Gottweis notes, the opening up of the German political system by the mobilizations and other political events of the 1970s meant that social movements operating in the 1980s, as the anti-biotechnology movement was, could employ a wider set of political strategies than were possible for movements in the past. The mass demonstrations, civil disobedience, and direct actions used so effectively by the anti-nuclear and other social movements of the 1970s were increasingly complemented by efforts to use the political and the legal systems to achieve change. Reflecting this new context for organizing, anti-biotechnology activists sought to use their newly established access to the Bundestag and the government's alleged commitment to greater openness and transparency to influence national policy toward biotechnology.[10] One concrete example involved the use of a parliamentary "enquiry commission," which could be set up to research and prepare detailed policy recommendations on complex political issues for members of the Bundestag. In 1984, the Green Party allied itself with the German Social Democrats to request that an enquiry commission be set up to consider the opportunities and risks associated with genetic engineering. After the Bundestag agreed to do so, the Greens created and funded a critically minded working group of medical personnel, natural and social scientists, and NGOs to advise them on the topic.

Although the Greens were unable to control the outcome of this three-year study, they did manage to widen the parameters of the national discussion around biotechnology. The final report of the Enquiry Commission, which consisted of several hundred pages of in-depth analysis and over 150 policy recommendations, was, on the whole, quite positive.[11] In response, Green Party critics submitted an extensive minority report (called a "special vote") to the commission and the media reflecting their deep disagreement. Their report offered a distinct social and political reading of biotechnology that emphasized the threats that modern biotechnology posed to society, rather than its promise. Its authors raised numerous objections to biotechnology and challenged the legitimacy of a purely science-based approach to regulation by suggesting that nonexperts had an important role to play in decision making.[12] Indeed, as the report's opening words suggested, "When dealing with a technology, pregnant with such far reaching consequences and affecting the very basis of life, it is essential to hold a wide ranging, public and essentially open debate on its development, before the development itself is allowed to proceed."[13] Furthermore, the minority report argued, such a technology should not be pursued at all unless its proponents could demonstrate a clear social need for it.

Although the majority of the Enquiry Commission's members took issue with the Greens' tactics and position, they decided to include the special vote in the commission's final report anyway, to avoid being accused of silencing the naysayers. The effect was to bring the activists' critique of genetic engineering into the national political space and to make public an alternative perspective that had been brewing quietly, in the context of small discussions taking place among environmentalists, feminists, and concerned scientists. As a result, biotechnology became the object of a growing public consciousness.

The tactics that activists used to challenge biotechnology's acceptance in Germany also included legal action, consistent with Gottweis's hypothesis of a strategic shift in German social movements' repertoires of collective action. In the 1980s, environmental activists utilized the courts to stop the German chemical company Hoechst from building a facility in Hessen to produce genetically modified insulin. Activists based their lawsuit on the argument that the government had not yet adequately proven the safety of biotechnology, and the building of such a facility would put the public in Hessen at risk, violating the government's responsibility to its citizens. The court took the side of the activists, holding up the plant's construction until a legal framework could be put in place that would formally regulate the technology.[14]

Under pressure from various segments of society, the German government took steps to create new legislation on genetic engineering, and in 1990 the first German Genetic Engineering Act was passed. As was the case with the final report of the Enquiry Commission, critics of biotechnology were unable to influence the law substantially in a way that significantly limited the technology's development. But they did successfully pressure the government to concede to their demand for greater public participation in decision making. One gain they made related to the number of nonbiologists who would sit on the Central Commission for Biological Safety (or the ZKBS, in German), the group that commanded regulatory authority over genetic research. Expanding the original twelve-member size of the ZKBS, which included eight biologists and one member each from the union sector, industry, environmental groups, and research organizations, the government agreed to add three more members, giving a stronger voice to ecologists and the environmental community. Environmental activists also managed to get two public hearing requirements included in the new Genetic Engineering Act, one of which related to the deliberate release of GMOs, and the other to the construction and operation of facilities for gene research. After these provisions were written into law, opponents of biotechnology used them to obstruct research experiments and to challenge the safety of GMO field trials. Three years later, however, this political opening closed up when the government repealed the public hearing requirements on deliberate release.[15]

In sum, although the anti-biotechnology activism of the 1980s did not profoundly change the German government's generally positive view and supportive policies toward genetic engineering, it did have a significant impact on the way biotechnology was perceived and regulated in Germany. Most important, activism politicized the issue, creating conflict around a technology about which there had formerly been little negative discussion.[16] Activists created this "dissensus," as Gottweis describes it, by discursively and politically challenging government and industry representations of biotechnology as unquestionably beneficial for the German economy and for society and by offering an alternative framing that focused on the risks and dangers it posed. As a result of these challenges, activists brought the existence of these new technologies into the public sphere and began to influence public opinion about them. Whereas most ordinary Germans had known little about the developments in molecular biology in the 1970s, the same was not true ten years hence. In the process, the German public was exposed to a far less sanguine interpretation of the technology than had been formerly presented to them. In addition, the impact of German activism traveled across the country's borders as German activists networked with other European activists, and the German Green Party carried its concerns into the European Parliament, where it helped to shape the European Commission's emerging biotechnology policy. Before we look at these transnational connections and the supranational political sphere, however, let us take an excursion into biotechnology politics in the United Kingdom, where a small group of activists were also beginning to organize around the issue.

"No Hormones in Our Milk!" The Struggle over bST in Great Britain

At approximately the same moment that German anti-biotechnology activists were challenging the supportive climate for biotechnology in Germany, a small cohort of British food activists, critical science and technology observers, animal welfare activists, and environmentalists were doing the same in the United Kingdom, albeit under a markedly different set of conditions.[17] Whereas the Green Party's ascendance into the Bundestag in the early 1980s gave German biotechnology critics a direct line of access to national politics and policymaking, British activists had no such access, through either the Green Party or any other. Furthermore, the conservative Thatcher government's antithetical stance toward regulation meant that Britain's early biotechnology critics would have to find other means of expressing their discontent than working through official government channels. Such an opportunity arose in the mid-1980s, when the U.S.-based biotechnology industry, led by Monsanto, began searching for a door through which it could walk its first major agricultural product into Europe. That product was bovine somatotropin (bST), a genetically engineered growth hormone that stimulates cows to produce more milk, as previously mentioned. According to Tim Lang, director of an independent research group called the London Food Commission (LFC), the industry's ill-conceived efforts to introduce bST into Europe gave activists their first opportunity to confront both the industry and the technology directly.

Lang had been keeping his eye on developments in the biotechnology sector for several years and had identified genetic engineering as one of the key issues on which the LFC should work. But when Lang was developing the LFC's initial strategic plan in 1984, there was not much for his public health and food safety group to organize around in the biotechnology arena. The industry was still young, no GM food products had come on the market, and the government was not making any effort to design regulations for what it perceived as a highly promising new sector. Sometime in 1986, however, a large brown envelope that appeared in the London Food Commission's mailbox altered Lang's political calculation. In it was a document prepared by a consortium of biotechnology companies that outlined the industry's strategy for moving into Europe. The document identified Britain as Europe's soft underbelly, the place where bST could most easily be introduced and the European country in which it was least likely to encounter resistance. "It was astonishing," exclaimed Lang, recalling his reaction to the industry's plan. If there was one thing British consumers did *not* need, in Lang's view, it was more dairy added into their already high-fat diets.[18]

After reading the industry's plan, Lang contacted a number of likely allies in other organizations to convey to them what was happening. One was Joyce D'Silva, a staff person at Compassion in World Farming (CIWF), who had learned about the U.S. Department of Agriculture's growth hormone experiments on pigs from Michael Fox, of the U.S. Humane Society, and was up in arms about genetic engineering. D'Silva was outraged at the idea that the industry had now designed a product that would artificially stimulate dairy cows to produce more milk and, in the process, subject them to a host of painful problems, including mastitis. She, Lang, and a few others formed a coalition that began to work on the issue. Their objectives were straightforward: to educate as many people as possible about bST; to convince government officials that bST was an unnecessary and harmful product for cows, consumers, and the dairy industry; and to prevent the technology from receiving British government and EU approval.

To achieve these goals, these bST critics utilized a variety of tactics. Lang and his colleagues established a broad-based work group to advise the London Food Commission's work on biotechnology and invited representatives from Britain's consumer organizations, farming groups, organized labor, and environmental groups to participate. They built up an alliance of small radical green organizations and larger, more "respectable" organizations, such as the Women's Institute, a long-established federation of farm women's organizations. Lang also asked Eric Brunner, an LFC colleague trained in biochemistry, to do a comprehensive study of bovine somatotropin. Brunner's report, titled *Bovine Somatotropin: A Product in Search of a Market*, reviewed the scientific evidence on bST, explored the animal welfare implications and potential human health effects, and analyzed the findings of an opinion survey the LFC had conducted on the technology.[19] Perhaps most important, the report challenged the industry's claim that bST was biologically identical to the natural hormone produced in a cow's own pituitary gland. In order to make their critique more visible, Brunner and several colleagues published a commentary in *Nature*, the most widely read British science journal, in 1994. Titled "Plagiarism or Public Health?" their article presented some of Monsanto's own data on bST and criticized the company's unwillingness to share these data publicly because of the problems they revealed.[20]

Through the contacts they had built up with their working party, the London Food Commission, CIWF, and other groups distributed the bST report and other papers and reports to the media, fellow activist organizations, and members of the British and European parliaments, exposing them to information they were not receiving from the industry. They also challenged biotechnology companies directly, asking tough questions, for example, at meetings Monsanto had organized with public health professionals, educators, farm groups, and other "decision leaders" around the country. D'Silva regularly wrote articles for the Compassion in World Farming newsletter, informing the animal welfare community of the struggle they were waging against bST and the biotechnology industry. At the end of each article, she provided a list of action suggestions: write your ministers of Parliament, contact the European Commission (the civil service for the European Union), call the dairy industry and food retailers, and *tell them all that you do not want this technology!*

Lang, D'Silva, and others traveled to Brussels personally to present dossiers full of critical scientific data and to share their viewpoints with the staffs and members of the European Parliament (MEPs). There, they joined efforts with activists from other EU countries who were also opposed to the technology. These critics of bST had surprisingly good access to Parliament, which was still finding its sea legs as a new pan-European political institution. Indeed, as a result of this intensive lobbying and the many scientific questions that NGOs raised, the European Parliament insisted that more studies be done before it would approve the growth hormone.

Ultimately, the activists were so effective in challenging bST that the industry's first major product never received approval in the European Union. But the broader significance of this and the other mobilizing efforts taking place at the time was far less obvious, lying buried beneath the surface. First, as food, animal welfare, environmental, and consumer activists began to work together within and across borders, they established the contacts and formed the personal relationships that would enhance their ability to work together in the future. Second, in the process of communicating their concerns about bST to a wide range of interest groups within Britain (recall the "unholy alliance" the London Food Commission created), these activists planted the seeds for a future revolt against biotechnology by the British population. They did this by publicly posing some seemingly simple but fundamentally heretical questions: What is this technology for? Who is going to benefit? What are the risks, and is there a good reason why society should agree to bear them?

Organizing at the European Union

In the late 1980s, activists initiated two other organizing efforts aimed at stopping biotechnology's development and deployment in Europe. One was the "No Patents on Life" campaign, whose goal was to prevent the European Parliament's approval of a new and industry-supported European directive on the legal protection of biotechnology inventions. The other effort focused on EU regulatory policy and sought to ensure that the European Union would enact stringent regulations on genetic engineering and the biotechnology industry. In both cases, these efforts were carried out by transnational networks of activists who inserted themselves into the EU's political process.

The Struggle over the European Patent Directive

In 1988, the European Commission proposed a piece of legislation that would strengthen the continent's intellectual property rights for industry investments in biotechnology.[21] Officially called the European Directive on the Legal Protection of Biotechnological Inventions, the purpose of the Patent Directive was to "harmonize" the patent laws of EU member states and to match the standards of legal protection that existed in the United States and Japan.[22] At the time, patent protection on the continent was ensured by two different systems, neither of which had its basis in EU legislation.[23] By establishing legislation at the EU level, the European Commission sought to assure biotechnology developers not only that they would enjoy strong and clear intellectual property protection in Europe, but also that this protection would be consistent across European countries. With a uniform and strong property regime in place, European biotechnology firms would be better able to compete with firms in countries where the biotechnology industry was developing most rapidly.

Even before the legislation was officially published in draft form in October 1988, opponents of the "Life Patents Directive," as it came to be known, started to organize against it.[24] In the spring of that year, a European group called the ICDA Seeds Campaign got hold of an advanced copy of the directive and began educating people about what it believed were the directive's profoundly problematic implications. They saw the directive as part of an alarming new trend toward life patenting that had started with the U.S. Supreme Court's decision in *Diamond v. Chakrabarty* and was spreading to Europe. In their view, the directive represented a major expansion of European patent law to cover a much broader array of living organisms (including whole groups of plants and animals), biotechnological processes, and information.[25] It would make genes into the "currency of the future" and give industry control of the whole supply chain, from basic genetic material to the products that make use of those genes and gene sequences, as well as future generations that carried that genetic information.

In June 1988, the group organized a workshop in Denmark around the Patent Directive and the general issue of life patenting. Seventy people from four dozen NGOs in twelve European countries attended, including a number of Green Party members. By the close of the event, workshop participants were so disturbed by what they heard that they decided to mount a huge effort to defeat the directive. The effort took the form of the "No Patents on Life" campaign, which would rage on for the next ten years.

The No Patents on Life coalition attempted to defeat the directive by mobilizing strong public pressure on individual members of Parliament and working through the Green Group in Parliament to propose a multitude of amendments to the draft directive.[26] Ultimately, the coalition's most powerful objections revolved around ethical issues, including the philosophical and moral implications of patenting (and thus claiming private property rights over) human genetic material and medical treatments (in the form of germ line gene therapies). Activists also contended that the directive would stimulate "biopiracy," a practice in which firms from the global North obtain genetic material from the global South and then sell it back to those farmers in the form of patented products. This, they argued, would end the free exchange of knowledge and resources that had historically undergirded the development of affordable health remedies, medical discoveries, and agricultural science.

In 1995, largely because of many parliamentarians' unease with these ethical issues (particularly those having to do with the patenting of human genetic material and germ line gene therapies), the Parliament rejected the draft patent directive in a 240 to 188 vote. Not surprisingly, those in the No Patents on Life coalition were ecstatic. As one NGO article on the battle joyously declared in its title, "The Directive Is Dead."[27] Two years later, however, the European Commission brought a revised directive back to the European Parliament for reconsideration. The new version included strong language that guarded against the patenting of human gene therapies and provided some limited protection to farmers engaged in seed saving (though not for commercial purposes). With its most serious concerns now addressed, the Parliament passed the Patent Directive. By and large, the pro-biotechnology forces had won. This was not to be the case with the second major struggle that activists waged in the EU, around regulatory policy. In this case, the outcome would turn out to be more favorable to the activists.

Regulatory Politics in the European Union

As suggested earlier, to most European government officials in 1980, biotechnology represented one of the promising new information technologies that was going to transform the ways in which societies worked, people interacted, and economies functioned. While few actual products had yet to come out of the industry, research on genetic engineering was moving fast, and it was only a matter of time before new applications would hit the marketplace. This arena was thus one that cried out for development of a formal, and forward-looking, policy.

As Sheila Jasanoff suggests, primary responsibility for developing a regulatory policy for biotechnology could logically have ended up in the hands of two different European Commission Directorates-General, the EU's central policymaking bodies.[28] One option would have been to place the responsibility in DG XII, the Directorate-General for Science, Research, and Development, whose jurisdiction includes the promulgation of scientific policy. Another was to place it in DG XI, the Directorate-General for the Environment, whose jurisdiction includes consumer safety and nuclear safety. Of the two, and true to its name, DG XI was known to be more open to input from environmental and consumer groups in European society. In either event, the task facing the relevant directorate was to create a policy that would govern the treatment of biotechnology on an EU-wide basis and that the various member states would be obliged to implement and use as a basis for harmonizing their own domestic policies.

After an interagency power struggle over who should take the lead in developing biotechnology policy, the Directorate-General for the Environment (DG XI) emerged as the victor, offering an important political opening for those concerned about biotechnology. Anti-biotechnology activists quickly seized upon this opportunity and began lobbying for a regulatory system that would recognize the uniqueness of the production process used to create GMOs.[29] Whereas the biotechnology industry and many EU government officials (particularly those associated with DG III and DG XII) strongly preferred a system that would regulate biotechnology on the basis of the food, medicine, and other products the technology produced (a "*product-based* system"), opponents of the technology, as well as many environmental officials, wanted a *process-based* system that would apply to all genetically modified organisms, whatever their final form and use. Under the former approach, some GMOs might not be subjected to any regulatory process or requirements if they were considered to be "substantially equivalent" to foods or crops currently on the market; under the latter approach, every use of genetic engineering techniques would be regulated by a new set of policies designed specifically for this purpose.[30]

Again, the Green Party served as the movement's chief conduit into the policymaking process. Although Green Party members did not have a place on the committee designing the most important part of this policy—Directive 90/220, pertaining to the deliberate release of genetically modified organisms into the environment—they actively offered their input to DG XI and strongly supported a process-based regulatory system.[31] For example, several Green ministers of the European Parliament played roles as interlocutors for the anti-biotech movement, and the Green Party's chief biotechnology policy adviser in Brussels interacted directly with those drafting the directive.[32] Given the significance of Directive 90/220, all interested parties made their positions well known, of course, including the Directorate-General for Science, Research, and Development (DG XII), which strenuously fought against a process-based system. But when all was said and done, the content of the Deliberate Release Directive that emerged from the Directorate-General for the Environment addressed the activists' primary concern. In the final analysis, if genetic engineering was to be permitted at all on the continent, then environmental and consumer groups and other critics got what they most wanted: a process-based system of regulation, which took as its starting point the basic GMO/non-GMO distinction.[33] As we illustrate below, once the European Commission issued Directive 90/220 in 1990, activists faced a new array of political opportunities for disrupting the biotechnology industry's forward march within Europe.

New Conditions for Organizing: Post-1995 Western Europe

At the end of the 1980s, the first genetically engineered food products started weaving their way through the U.S. regulatory process to become a commercial reality. The first product, approved by the U.S. Food and Drug Administration in 1989, was chymosin, a genetically engineered bacteria used in cheese making. This was followed by bovine somatotropin, the approval of which had been effectively halted in Europe, as we have already seen. Then in 1996, the British company Zeneca successfully introduced a genetically engineered tomato paste in order to test public acceptance of the technology. By the end of that year, however anti-biotech activists had galvanized an increasingly visible and consumer-oriented resistance to the technology. Clearly, conditions for organizing had changed.

The trigger was a new group of commodity crops engineered to resist the toxic effects of specific herbicides or to include a gene that functioned as a pesticide (*Bacillus thuringiensis,* or *Bt*) or both.[34] The most important of these were Monsanto's Roundup Ready soybeans and *Bt* corn, which received U.S. Department of Agriculture approval in 1995 and EU approval in 1996. Not only did the commercial deployment of these crops create a new group of commodities around which to organize, but also the U.S. firm introducing them was seeking to do so without the consumer's knowledge of their genetic modification. What is more, these two commodities (soy and corn) were processed into so many different food products that they would become unavoidable elements of people's diets.

Perhaps the person to recognize the propitiousness of the moment most explicitly was Benny Haerlin, the former German Green MEP who had been working to stop biotechnology for a decade. In the summer of 1996, Haerlin was coordinating an anti-toxics campaign for Greenpeace

International when he received a call from an executive at an upscale German supermarket chain called Tengelmann's. The executive told Haerlin that he knew that a shipment of GM crops was scheduled to arrive in Europe later that year and wanted to know whether "Greenpeace [would] have a problem with that." Although Greenpeace did not have an anti-GMO food campaign in place at the time, Haerlin answered in the affirmative, hung up the phone, and began to organize one.[35] Haerlin convinced Greenpeace that this was a crucial moment to push the GMO issue and persuaded the organization to assign fifteen full-time organizers to the issue. When the ships carrying the crops arrived in European ports, Greenpeace activists were ready and waiting. They surrounded the ships in a small flotilla of boats and unfurled large banners calling for a ban on GM food imports. Members of the press corps were on the shores, snapping away with their cameras.

Between 1996 and 1999, movements against genetic engineering gathered steam all across Europe. Growing fears of the environmental and social implications of a poorly regulated and uncontrollable technology created a sense of common cause among a wide range of environmental organizations, agricultural organizations, and public interest groups. In Britain, the Royal Society for the Protection of Birds became engaged because it feared the loss of vital bird habitats if GM traits should make plants unpalatable or stray across species. In France, environmental groups such as Ecoropa-France and France Nature Environment got involved because they were worried about GMOs' impact on the environment. They were joined by a group of French scientists, mainly population biologists, who felt that their scientific expertise was ignored when the EU authorized an herbicide-resistant GM canola in 1996.[36] Farming organizations such as the British Soil Association, which represented organic farmers, and Confédération Paysanne, the French peasants' confederation, took up the issue because they worried that GM crops could destroy organic farming and feared that this new technology would harm the livelihoods of small-scale farmers. To the French anti-globalization group Attac, GMOs represented a clear, negative manifestation of globalization.[37] Both Greenpeace and Friends of the Earth–Europe established major transnational campaigns against GM food in 1996, thanks to increased funding from several concerned foundations, including the Goldsmith Trust.

The boom in activism was fueled by popular concerns over food safety and consumer right-to-know issues, which were gaining salience as a result of food and health scares that had rocked the continent. The most significant of these, of course, was mad cow disease and its human variant, Creutzfeldt–Jakob disease, which were reaching their peak as animal and human health disasters in Britain and beyond. Creutzfeldt–Jakob disease had a particularly devastating and terrifying impact in the United Kingdom, leading to almost seven hundred human deaths in the 1990s.[38] The identification of AIDS-tainted blood in France, the discovery of dioxin in animal feed, and a rash of illnesses associated with Coca-Cola added to the European public's sense of vulnerability. In this context, it is not surprising that citizens would be frightened by any large-scale tinkering with their food supply and would insist on the right to know what they were consuming.

Movement Strategies after 1995

Reflecting the new blood and energy flowing into the European anti-biotech movement and the crucial fact that GM food was now a commercial reality, European activists broadened their repertoire of contentious actions. The more theatrically oriented arm of the movement, drawing on its experience with environmental and anti-nuclear activism, stepped up its use of symbolic protests. In one eye-catching stunt, a group of Greenpeace activists slipped into the 1996 World Food Summit in Rome and stripped off their clothes to reveal naked bodies brightly painted with anti-GM slogans. Another group of activists dressed themselves up as "Super Heroes against Genetics" and took over the headquarters of Monsanto-UK, adorned in capes, tights, and underpants.[39] In early 1999, Greenpeace dumped over four tons of genetically modified soybeans on the doorstep of Prime Minister Tony Blair's official residence from a large truck decorated with a banner reading, "Tony, Don't Swallow Bill's Seed."

Movement members also engaged in an increasing number of direct actions. They entered supermarkets and prominently labeled foods as genetically modified. They set up squatter camps on plots planted with GM crops to draw attention to the issue. And they destroyed GM trial crops in farmers' fields. In one well-publicized case of "bio-sabotage," protestors illegally pulled up field trials of a GM corn crop that had been planted close to an organic farm in Totnes, England. The British Soil Association and other groups organized in support of the arrested protestors and the farmer who was taking the government to court for the GMO trials, and they used the legal system and the media to turn the case into a cause célèbre.[40] Lord Peter Melchett, executive director of Greenpeace UK, organic farmer, policy director of the British Soil Association, and one-time Labour minister for the environment, became one of the most visible faces of the movement when he was arrested, charged, and acquitted of criminal damage for ripping up a GM maize crop. In 1999, French activists began to destroy GM field trials in public acts of defiance that Confédération Paysanne leaders Jose Bové and Gilles Luneau likened to those carried out during the 1773 Boston Tea Party.[41]

Activists repeatedly blew the whistle on the introduction of unauthorized GMOs in order to draw attention to weak government controls on the technology. For instance, in 1996, Austrian environmentalists publicized the fact that their government had released GM potatoes without the required approvals.[42] In another noteworthy case, activist organizations called for the destruction of an entire contaminated crop when Advanta Seeds UK admitted that GM rapeseed had inadvertently been mixed with conventional seed imported from Canada and sold in the United Kingdom and Europe for two years. They then publicly berated the British Food Standards Agency for failing to act as a public watchdog.

Activists combined these actions with new organizing efforts focused on pressuring food manufacturers and retailers to stop using and selling GM foods. In the process, they took direct aim at an unforeseen Achilles' heel of the industry: downstream consumers. Friends of the Earth–Europe, Greenpeace, and the Women's Environmental Network in Britain, among other groups, organized sophisticated supermarket campaigns, urging consumers to contact supermarket managers about their concerns regarding GMOs. As part of a "good cop, bad cop" strategy, Friends of the Earth–Europe maintained a public scorecard of retailer behavior, publicly congratulating companies that made decisions they favored and publicly chastising those that were reluctant to budge on the issue. In short, they played one retailer against another in an effort to force the major supermarket chains to reject GM food. These campaigns put European supermarkets on the defensive and ultimately led them to stop selling GM food.

To publicize their opposition to GMOs, activists energetically courted the media, which often gave them sympathetic coverage. The British press and the BBC extensively covered the destruction of GM field trials in the UK countryside and the court cases that typically followed these acts. In 1996, the French daily *Libération* ran a front-page article headlined "Beware of Mad Soya!" picking up on an activist refrain and drawing broad public attention to the issue.[43] In Denmark, the media treated biotechnology very negatively, presenting it, in the words of one academic research team, "as an economic project that will only benefit biotechnology companies while posing risks to the environment, consumers, the ecology, and animal and human health."[44] Throughout Europe, the term *Frankenfoods* became the media's shorthand for GMOs. The media's receptiveness was a real advantage to activist groups, because it enabled them to keep their momentum high during ongoing campaigns and direct actions. In addition, extensive media coverage enabled the movement to generate and sustain pressure on governments and to keep the issue in the political spotlight.

Discursive Challenges

A final weapon the movement mobilized against the industry and the technology, and one that crosscut all of the tactics just described, was a discursive one. As noted in chapter 3, members of the anti-biotech movement had a worldview that differed dramatically from that held by biotechnology industry officials and, for that matter, from that of most policymakers. This worldview stressed people's democratic right to take part in the decisions affecting their lives and the dangers of an unbridled faith in science and technology. It rejected a utilitarian perspective on nature (that is, one that viewed living things and their genes "as one big Lego set" that could be rearranged to serve human purposes better) and the notion that life itself should be subjected to private property claims. Seeing the world from this vantage point, activists advanced a set of counterdiscourses that directly challenged the discourses of their opponents.

One of the most powerful discourses the movement mobilized involved the unknown health and environmental risks associated with the reengineering of plant and animal (including human) genomes. Whereas proponents of agricultural biotechnology claimed that genetic engineering offered increased precision in plant and animal breeding, activists vociferously challenged this assertion and stressed the uncertainty of the science, arguing that much was still unknown about the workings of biological organisms and that these new gene-manipulating technologies could lead to health and environmental problems that even the best "scientific experts" could not foresee. In a particularly strategic move, activists discursively linked the continent's horrifying experience with mad cow disease to the understudied and unknown risks posed by genetic engineering, which certain governments had also proclaimed was "perfectly safe." They used terms such as *genetic pollution* and *genetic contamination*, in addition to *Frankenfoods*, to describe GMOs and to emphasize the risks of gene flow, biodiversity loss, and the notion that genetic engineering was a fundamentally uncontrollable technology.[45] In a context in which the public had a strong sensibility toward the environment and little trust in government regulatory authorities and industry, these discourses resonated strongly.[46]

Another discourse the movement advanced invoked the idea that accepting GM crops would lead to an inevitable transformation of European agriculture, changing farming systems into something that resembled U.S.-style agriculture. Activists portrayed agricultural biotechnology as the latest trend in large-scale, industrial agriculture, one that carried the potential to destroy the thousands of small farms that dotted the European countryside. To many Europeans who felt a strong affinity for rural life and the countryside, this idea was deeply offensive. This discourse was particularly persuasive in France, where artisan agriculture and the notion of *"terroir"* were part and parcel of people's food identity and culture.

Profoundly disturbed by their governments' overt enthusiasm for biotechnology, the movement members also promoted a set of critical discourses about the state and its responsibilities with respect to the new biotechnologies. Anti-GMO activists harshly criticized government officials for rushing to support the biosciences and the biotechnology industry, rather than seeking to regulate them stringently. Pointing to the spate of food-related public health scares and disasters that had swept through the continent in the 1990s, they questioned whether government policymakers were doing their jobs of protecting the public or were mainly acting as shills for industry. Furthermore, activists vigorously attacked their own governments and EU institutions for not doing enough to secure the public's right to know and choose what it was eating. All across Europe, the anti-biotechnology movement argued that consumers had a basic right to know what was in their food and to be able to make their consumption choices on that basis. These discourses resonated with large portions of the European electorate who were already feeling politically disenfranchised by the creation of the EU, with its supranational forms of governance. Ultimately, such discourses had a powerful effect on the policy sphere, where they helped to compel both individual European governments and the European Union to develop policies that would ensure the "right to know" through GMO labeling.

Cultural Misreadings and Marketing Missteps

During the latter half of the 1990s, Monsanto made a number of critical missteps that helped the activists stir up anti-biotechnology sentiment among the European populace. Monsanto had the misfortune of possessing all of the features that Europeans associate with "the ugly American": arrogance, cultural insensitivity, and a deeply held belief that "our way is better." In a manifestation of the corporate culture described in chapter 2, Monsanto stormed into Europe like a general going to war, making one cultural and political gaff after another in its dealings with the European public and governments. It also profoundly misread European public opinion and its political significance.

Perhaps the most glaring example of Monsanto's errors in judgment was the company's decision to send its genetically modified crops to Europe unlabeled, despite being explicitly warned not to do so by members of the European biotechnology industry. In the summer of 1996, an executive from the British biotech company Zeneca, Simon Best, flew to St. Louis to meet with Monsanto's chief executive officer, Robert Shapiro, to advise him about how the company should enter the European market. Drawing on his own experience with introducing a GM tomato paste into British supermarkets, Best strongly urged Shapiro to label Monsanto's Roundup Ready soy before shipping it to Europe and to advise consumers well ahead of time that it would be coming. At harvesttime, however, Shapiro decided he knew better than this British executive and sent the GM soy to Europe unannounced and unlabeled.[47] His decision infuriated anti-biotechnology activists, who saw Monsanto's refusal to label as indicative of the company's complete disregard for their democratic rights as consumers to know and to choose what they were eating.[48] It also reflected Monsanto's misreading of public sensibilities around food safety, which had been greatly inflamed by the food and health crises mentioned earlier. Not surprisingly, activists strategically used Monsanto's behavior as grounds for criticizing the company for being hostile to consumers' desires.

Monsanto made another poor judgment call in its decision to embark upon a $1.6 million advertising campaign in 1998 designed to "educate" the British public about biotechnology's many benefits. In its advertisements, the company made far-reaching claims that activists were able to attack vigorously and effectively as specious, including the claim that world hunger would be reduced through the use of biotechnology.[49] "Worrying about starving future generations won't feed them. Food biotechnology will," read the headline on one of Monsanto's full-page advertisements in several national newspapers, for example. Yet as Tony Juniper, from Friends of the Earth–UK, noted in a National Public Radio interview, "People know over here [in Britain] that in fact poverty is the principal cause of hunger in the third world; there isn't a shortage of food. And people are, quite rightly in our view, interpreting these claims as a means to get bigger markets, to get a return on the investments that Monsanto has put into generating these new crop technologies."[50] Whether or not Juniper was right about the UK public's understanding of the root causes of world hunger, he did identify a sentiment that much of the British public apparently shared, which was that Monsanto and other biotechnology

corporations were in the business of biotech to make money, not for altru-
istic reasons. Trying to pretend otherwise was counterproductive.

Monsanto's misreading of the political and cultural climate in Europe
also led it to misjudge the likely impact of its efforts to pressure EU officials and those of various individual European governments into speedily
approving its products. Monsanto took an aggressive lobbying approach
in Europe, directing employees in its European offices to work as closely
as possible with the various agencies and policymakers involved in making regulatory decisions about GMOs. Company executives also made
their feelings known through the U.S. office of the trade representative,
headed at the time by Mickey Kantor, a close friend of Robert Shapiro,
then Monsanto's CEO.[51] Given the increasingly sensitive nature of the
biotechnology issue, however, these tactics backfired. From most policymakers' perspectives, Monsanto's pressure tactics were not only annoy-
ing but also failed to recognize the company's lack of maneuvering room
given the sorts of public pressure they were under.

Collectively, these behaviors and political miscalculations made Monsanto into the perfect target for activists, enabling them to vilify the firm
and the technology simultaneously. They also fueled growing sentiments
of anti-Americanism among European consumers and allowed the anti-
GE movement to portray Monsanto successfully as the agent of American
food imperialism, the devil incarnate who was trying to shove a dangerous technology down people's throats, whether or not they wanted it.
Several years later, Robert Shapiro openly admitted that his company had
made a grave mistake in its handling of its European market entry. "Our
confidence in this technology and our enthusiasm for it has widely been
seen, and understandably so, as condescension or indeed arrogance," he
said in a public broadcast at a Greenpeace-sponsored conference in 1999.
"Because we thought it was our job to persuade, too often we forgot how
to listen."[52]

Impacts on Public Opinion

Largely as a result of the activists' actions, Monsanto's missteps, and
the massive press coverage the issue received, Europeans' awareness of
GM food increased markedly after 1996. As activist discourses resonated
with the public, public opinion began to turn more firmly against GMOs.
Whereas the vast majority of Europeans were generally agnostic about
agricultural biotechnology in the early part of the decade, "widespread
public ambivalence about GM foods . . . [gave] way to widespread public

hostility" by the decade's end.[53] Between 1996 and 1999, the fraction of the public opposed to GM food rose 20 percentage points or more in Greece, Luxembourg, Belgium, and Britain, with changes in Portuguese, French, and Irish public opinion not far behind (see Table 2). By 1999, only one-fifth of western Europeans were strongly supportive of GM food, and only a third were supportive of GM crops.[54] When asked whether they agreed with the statement, "If anything went wrong with GE (genetically engineered) foods, it would be a catastrophe," fully 57 percent of a large sample of Europeans answered yes.[55]

The shift in British and French public opinion was especially marked. According to a private study commissioned by Monsanto, British attitudes toward GM foods grew much more negative over the course of 1997 and 1998, rising from 38 percent to 51 percent of the population.[56] By the summer of 1998, only 14 percent of the British public was found to be "happy" with the introduction of GM foods, and 96 percent wanted them labeled.[57] Helping to push British consumers away from biotechnology was a widely read letter published by Prince Charles in the London *Daily Telegraph*. In it, this sympathetic elite noted that he personally would not knowingly eat GM food or feed it to his family. "I happen to believe that this kind of genetic engineering takes mankind into realms that belong to God and God alone," the prince penned.[58]

In France, distaste for agricultural biotechnology also exploded. Whereas fewer than half of the French public (46 percent) were opposed to biotechnology in 1996, that figure rose to almost two-thirds (65 percent) in 1999 (Table 2). An index measure of "biotechnology optimism" generated by researchers at the London School of Economics reflected the same sentiment, falling 21 percentage points over this same period (from 46 in 1996 to 25 in 1999) for the French sample.[59]

The Road to Market Closure

Changing public attitudes toward GM foods in Europe, together with a growth in concern about food safety, significantly enhanced activists' efforts to prevent GM food from becoming commonplace in Europe in the latter half of the 1990s. The outcome of market closure, with which we started this chapter, occurred in two mutually reinforcing ways. One involved the food retailing industry's decision to go GMO free, and the other involved a political shift at the level of the European Union, ultimately resulting in a moratorium on new GM crop approvals. Anti-biotechnology activists played a crucial role in both of these processes.

Table 2. Changes in European public attitudes toward genetically modified foods, 1996–99

Country	Percentage of population opposed, 1996	Percentage of population opposed, 1999	Percentage point change
Austria	69	70	1
Sweden	58	59	1
Denmark	57	65	8
Norway	56	65	9
Greece	51	81	30
France	46	65	19
Germany	44	51	7
Luxembourg	44	70	26
Italy	39	51	12
Britain	33	53	20
Belgium	28	53	25
Portugal	28	45	17
Ireland	27	44	17
Finland	23	31	8
Netherlands	22	25	3
Spain	20	30	10

Sources: Adapted from Gaskell, Allum, Bauer, et al. 2000, Table 5; data from the 1996 and 1999 Eurobarometer surveys on biotechnology.

European Supermarkets and Food Processors Defect

As noted above, one of the anti-GMO movement's chief strategies after 1995 involved organizing pressure campaigns on European food retailers. In March 1998, these supermarket campaigns began to pay off when a maverick frozen food company named Iceland Foods agreed to renounce the use of GM ingredients in its store brand products. Over the next year and a half, dozens of other European food companies followed Iceland's lead and moved to clear their own shelves and brands of GM foods (see Figure 3). Among them was virtually every major supermarket chain and food manufacturer on the continent as well as the British Isles. Iceland Foods had set off a chain reaction.

Several factors facilitated the efficacy of the movement's consumer-oriented activism and led retailers to go GM free. The first was the organization of the processed-food commodity chain in which the agricultural biotechnology industry is situated. As we explained in the Introduction, agricultural biotechnology firms do not sell their products (that is, GM seeds) directly to food consumers but, rather, sell to farmers. Farmers sell their crops to grain elevators or handlers, who then sell the milled grain to food processors. Food processors' products are then sold to supermarkets, the fast food industry, and restaurants, which sell them to final food consumers. Furthermore, unlike the output of industries that produce products for a wide range of uses and markets, the main output of the agricultural biotechnology industry (GM seeds) is used to produce food for human consumption. Consequently, even though final food consumers are not the *direct* customers of agricultural biotechnology companies, activists could exert consumer pressure at the downstream end of the food commodity chain to harm the firms at the upstream end, that is, the biotechnology industry. By giving GMO foods a bad name and mobilizing consumers to express their concerns to retailers and processors, which they did in droves, this is precisely what activists did.

The structure of the food retail sector significantly augmented the impact of the movement's retailer-oriented activism. Across Europe, the supermarket sector had experienced a significant process of growth and concentration during the 1980s and 1990s and had come to be dominated by a relatively small number of large and influential firms.[60] Competition within this sector was extremely fierce and rested on these firms' abilities to establish themselves as purveyors of competitive pricing and quality, which was captured in their house brands. In this highly competitive environment, any significant customer defection posed a serious threat. This made supermarkets an excellent target for activist attacks, particularly when those attacks were aimed at questioning the quality of a firm's store brand. Thus, supermarkets occupied a position in the commodity chain that was both very powerful and very vulnerable.

The desirability of food processors and retailers preemptively rejecting GMOs was bolstered both by the public's growing fear about the safety of the European food supply and by its increased interest in the quality of the food consumers were buying. In the United Kingdom in particular,

Supermarket chains

March 18, 1998	Iceland	Will no longer use GM ingredients in own brand products
February 5, 1999	Carrefour	Will no longer use GM ingredients in own brand products
February 13, 1999	ASDA	Will no longer use GM ingredients in own brand products
March 15, 1999	Marks and Spencer	Will no longer use GM ingredients in own brand products
March 17, 1999	Sainsbury's	Will no longer use GM ingredients in own brand products
March 18, 1999	Co-op	Will no longer use GM ingredients in own brand products
March 18, 1999	Waitrose	Will no longer use GM ingredients in own brand products
April 27, 1999	Tesco	Will no longer use GM ingredients in own brand products; will label all other products that contain GMOs

Food processors

April 27, 1999	Unilever UK	Will no longer use GM ingredients in its products in Britain
April 28, 1999	Nestlé	Will phase out GM products as soon as possible
April 29, 1999	Cadbury-Schweppes	Will phase out GM products as soon as possible
May 7, 1999	RHM	Will stop using GM corn and soy in its products
May 8, 1999	Northern Foods	Will ban GM ingredients
June 16, 1999	Hazelwood Foods	Will eliminate all GM ingredients from frozen food by year's end

Figure 3. Partial list of major European food processors and supermarket chains to go "GMO free." Sources: BBC News 1998, 1999a; Waugh 1999; Lean 1999a, 1999b; Roberts 1999.

the public questioned not only whether the foc
whether the government could be trusted to e
of that supply. After all, the British governm(
industry to feed cows with the ground-up r(
cluding diseased ones, which led to the horrifi
form encephalopathy, or mad cow disease.
the source of mad cow disease to this indu:
authorities continued to insist that the publi(
consuming British beef on the grounds that the disease was ᴄᴏ…
animals. Months later, scientists discovered that the disease had jumped
species and was appearing in humans who had eaten British beef.[61]

In such a climate, the food industry as a whole became extremely
vulnerable to the perception that the food it was selling was unsafe. Food
retailers and processors were thus inclined to do whatever was necessary
to maintain the public's confidence and trust, because their company's
brand names and reputations were on the line. As the president and CEO
of Novartis's Gerber baby food division noted in explaining why his company decided to go GMO free, "I have got to listen to my customers. So,
if there is an issue, or even an inkling of an issue, I am going to make
amends. We have to act pre-emptively." The company's vice-president for
research, Jan Relford, was even blunter: "The parents trust us; if they
don't trust us, we are out of business."[62]

Finally, the willingness of the European food retailing sector to desert
the agricultural biotechnology industry and its large-scale investment in
GMOs was enhanced by the unequal relations of dependency that existed
between the two groups and by the failure of the U.S. agricultural biotechnology industry to consult European food manufacturers and retailers
about their willingness, and ability, to sell GM food. Reflecting its U.S.-
centric and supremely confident attitude, Monsanto had arrived on the
continent without so much as a phone call to any major food processing
and retail company in Europe, even though the company was thoroughly
dependent on these firms to buy and sell its products. This dependence,
however, was not mutual, because European food retailers and manufacturers could survive perfectly well without getting involved in the GM
food trade. (After the mid-1990s, in fact, they were likely to be better off
if they stayed *out* of the GM food business, since their customers were
telling them that they did not want GM food.) The failure of the U.S.
biotechnology industry to persuade European retailers to "buy in" on the
technology turned out to be a serious error in judgment. As one activist
sardonically noted, "They just assumed that they could manage it."

5, European activists also took advantage of new political op-
ties associated with the EU's new regulatory system to inflict harm
e biotech industry. To understand what these opportunities involved,
e must first understand the process for obtaining approval to introduce
a GMO into the EU, as well as certain peculiarities of the EU system.

The GMO approval process entailed a number of steps. First, in order
to introduce a GMO into the market, the interested party had to submit a
dossier of information to the designated regulatory authority (the "competent authority") in one of the EU member states. In that dossier, the party
seeking approval was required to provide a detailed risk assessment of
the GMO.[63] If a competent authority (CA) was unsatisfied with the risk
assessment, it could return the dossier to the applicant with a request for
more information or simply refuse to approve the GMO. If, however, the
competent authority concluded that the GMO did not pose a threat, then
that CA would serve as the application's sponsor and send the dossier on
to the European Commission with a positive recommendation. The commission would then distribute it to all the other member states for their
approval. If no member state objected during the allotted time period, the
commission would refer the matter to the European Union Council (made
up of representatives of the member states) and to the European Parliament
for a vote. If a "qualified majority" of both bodies (that is, at least two-
thirds) voted in favor, the application was approved, and each member
state was then obliged to allow the GMO's use within its borders. If a qualified majority was not in favor, the request was sent back to the European
Commission, which would have its own scientific committees reevaluate
the dossier and then decide whether or not to approve it.[64] In general, com-
missioners and policymakers were inclined to approve the technology, both
because they were reluctant to impede what they saw as progressive change
and because they were eager to foster open trade among countries.

Given the multistage nature of this process, the scope of the scientific
risk data that could be requested under EU policy, and the requirement
that an application be approved by majority votes in both the European
Council and the European Parliament (which is democratically elected
and sensitive to public opinion), applications could become bogged down
at several points. For example, as specified in EC Directive 90/220, parties had to submit thirteen categories of technical data, many comprising
numerous subsets of information, as part of their information dossiers.[65]
Category 11, for example, requests information on the "pathological, ecological, and physiological traits" of the GMO, including:

(a) classification of hazard according to existing Community rules concerning the protection of human health and/or the environment;

(b) generation time in natural ecosystems, sexual and asexual reproductive cycle;

(c) information on survival, including seasonability and the ability to form survival structures e.g.: seeds, spores or sclerotia;

(d) pathogenicity: infectivity, toxigenicity, virulence, allergenicity, carrier (vector) of pathogen, possible vectors, host range including non-target organism. Possible activation of latent viruses (proviruses). Ability to colonize other organisms;

(e) antibiotic resistance, and potential use of these antibiotics in humans and domestic organisms for prophylaxis and therapy;

(f) involvement in environmental processes: primary production, nutrient turnover, decomposition of organic matter, respiration, etc.[66]

Indeed, not long after the first few petitions arrived at the European Commission and were approved by the European Council and Parliament, activists recognized the myriad possibilities of disrupting the approval process and began working with their allies among the member states to do so.[67] Austria, Denmark, Greece, and Luxembourg had all taken very cautious attitudes toward GMOs in their own countries and were motivated to create a strong set of biotechnology regulations at the EU level. (For instance, Denmark had passed its own highly restrictive GMO legislation in 1986, and Austria was among the first to ban crops that had been approved by other countries, on the grounds that unresolved questions remained about their potential risks.) Accordingly, these countries were quite willing to object to an application on risk-based grounds or to request additional data, or both, before agreeing to render their decision if they believed there was just cause.

Activist groups such as Greenpeace and Friends of the Earth–Europe worked hard to support these countries' competent authorities and to slow down the approval process for GMOs. They facilitated the access of these CAs to scientific studies that pointed to the risks of specific GMOs, they steered them toward scientists who could criticize the risk data that firms had provided, and they backed their efforts to interpret the Patent Directive broadly and to reject GMO applications on the basis of their implied risk.[68] As one activist described the strategy to us,

> [We had] a lot of very well-informed activists who read the rules and said, "These don't make too much sense," [and] who started exploiting the particular fragilities . . . in the European system. Because you know . . . we had, at that stage, fifteen member countries, all of whom

had to agree on these things. It was kind of nice if you happened to find two or three who said, "Hey, I'm not too sure." And [you could] play the game with that.

Later in our discussion, he offered some additional reflections on why this strategy worked:

The biotech industry was not too clever. The things they were applying for permission for were not particularly well-developed, not very well thought out. Their applications were really quite sketchy, so it was easy to critique them.

. . . The activists were more ahead of the game than the biotech companies were. They [the companies] just thought it was going to be OK. . . . They'll just send the papers in, and you know, nothing to worry about. So they were a bit taken aback when they started to get a lot of flack. Monsanto, in particular, . . . was demonized very heavily.[69]

As the politics around biotechnology heated up in Europe, more and more petitions for GMO approval became logjammed by requests for more risk-related data. By 1998, the approval system was so bogged down that no new applications were getting through. In part, this reflected the effectiveness of the activists' strategy of raising objections to the dossiers and a few naysayer countries' concerns about the technology. But it also reflected two other phenomena: namely, the shift in public opinion against GMOs and the significant pressure that a number of EU member states, particularly the United Kingdom, France, and Germany, had come under to become more responsive to the public's will.[70] In the context of biotechnology, this meant respecting the public's growing antipathy toward GMOs. As a result, explained the activist just quoted above,

in 1998 . . . the whole system became deadlocked. The regulatory committee, which was supposed to give advice, kept giving negative advice. The [European] Commission was supposed to overrule it. It didn't know what to do. And they were supposed to send it up to the ministers [of the European Parliament] for them to decide, but the ministers didn't want to take this hot potato either. And they all just sort of ground to a halt.[71]

These conditions created market uncertainties that wreaked havoc on the biotech industry, not least because farmers became reluctant to plant even those GMOs that *had* been approved.[72]

Adding to the strain on the industry was the decision by some countries to unilaterally ban the marketing of certain approved GMOs, under the protection of Directive 90/220's Article 16. This article allows an EU member state to provisionally restrict GMOs that the EU has approved if that state has "justifiable reasons to consider that a product constitutes a risk to human health or the environment" and if the state informs the commission in writing of its action and justifies its decision.[73] In 1997, Austria, Italy, and Luxembourg all invoked Article 16 to ban a GM corn that the commission had just approved, and in 1998, France did the same with respect to two GM oilseed rape varieties.[74] In effect, this article allowed countries not only to defect from EU-wide rules, but to do so in an insurrectionary and destabilizing manner.

In 1999, what had been an informal stoppage turned into a more formal moratorium on new GM crop approvals. The most immediate cause was a dramatic shift in the position of the French and British governments. Although France had been one of the most pro-biotechnology governments up through 1997, it began taking a more precautionary approach, prompted by increasingly loud calls from activist organizations, scientists, and the public, who were disturbed about the government's insufficient attention to risk. The British government was also forced to abandon its staunchly pro-biotech position as its public protested the democratic deficit in policymaking and called for greater transparency and public accountability on food issues.[75] These reversals were politically significant. In June 1999, France called for a suspension of all commercial GMO authorizations at the EU Council of Ministers' meeting. Four other countries supported this position, and seven adopted a separate declaration that also made an argument for greater caution.[76] The effect was to close European markets to new GMOs for the next four years. Even after the moratorium was lifted in 2003, very few GM crops were approved, reflecting the continent's continued opposition to genetically engineered food and seed products.

Reverberations of European Activism

Thanks to the global nature of the commodity chain for food, the impact of Europe's market closure to GM foods in the late 1990s was felt far beyond Europe's borders. The U.S. farm sector felt the squeeze most directly. In 1999, U.S. grain elevators started rejecting suspected GM corn and soy crops unless farmers could prove they were GM free. The high rate of GM crop adoption in the United States, combined with a system that lacked the capacity to segregate GM from non-GM grain, left North

American farmers and exporters extremely vulnerable to this market shift. Within a few years, the U.S. agricultural sector lost tens of millions of dollars as grain sales to Europe plummeted.

When the EU market closed its doors on GMOs, governments and publics in other countries also reacted. As we show in chapter 6, governments of poor countries became extremely sensitive to the potential economic ramifications of adopting GM crop technologies. For countries whose economies still depended very heavily on agriculture, particularly in Africa and parts of Asia, the risks of losing international markets because of GM crop and land "contamination" were perceived to be very high. Global South governments' concerns about GMOs were exacerbated by the fragility of their ecological systems. Thus, many countries started to impose regulations that would stop the movement of GM seeds into their countries.

Portions of the public in many countries, particularly among the elite classes, also began questioning the safety of a technology that Europeans were openly and publicly rejecting. If it was not considered safe by the Europeans, something must be the matter with it, they reasoned. Hence, even though there was no incontrovertible proof of any negative health effects caused by the technology, the idea that the scientific jury was still out and that serious problems could present themselves in the future traveled rapidly around the world, riding on the currents of the press coverage and the Internet. The European market shift and the myriad reactions it caused had a powerful dampening effect on the financial outlook of the industry. By 2000, few people continued to believe that agricultural biotechnology had a bright and promising economic future or was a smart place to invest one's money. The only country that remained very open to the technology and optimistic about it was the United States.

CHAPTER FIVE

Creating Controversy in the United States

NEW CONSUMER CAMPAIGN TARGETING KRAFT FOODS LAUNCHED IN 170 CITIES TODAY. Washington, DC—Today consumer activists in over 170 cities around the United States, Canada, and Australia launched a new campaign that calls on Kraft Foods to remove untested, unlabeled genetically engineered ingredients from its products. . . . The Genetically Engineered Food Alert coalition demonstrated at grocery stores around the country . . . to draw attention to the public health and environmental concerns associated with genetically engineered foods and to inform consumers that Kraft Foods' genetically engineered products are neither adequately safety-tested nor labeled.

—GE Food Alert Press Release, February 6, 2002

At the turn of the millennium, anti-GMO activism was at an all-time high in the United States. In mid-2000, as we noted in the Introduction, *Time* magazine ran an article that pointed to "a carefully coordinated start of a nationwide campaign to force the premarket safety testing and labeling of . . . genetically engineered organisms" as part of which "seven organizations . . . were launching the Genetically Engineered Food Alert, a million-dollar, multiyear organizing effort to pressure Congress, the Food and Drug Administration and individual companies." In the fall of 1999, activist pressure had forced the U.S. Department of Agriculture (USDA) to hold public hearings on GMOs in three major cities across the United States. Exactly a year later, a group of Washington-based activists exposed the fact that a variety of GM corn approved only for animal consumption, known as Starlink corn, had made its way into the nation's food supply and corn shipments to other countries. Nightly news programs broadcast story after story about contaminated taco shells in U.S. supermarkets and the fallout from the illegal Starlink exports. What started out as just an embarrassing scandal ended up costing the industry and the USDA hundreds of millions of dollars. Public awareness of genetic engineering increased markedly, and the industry was thrown on the defensive. These growing rumblings of civic action contributed directly to the sense of crisis that we observed at the 2001 conference of the Food and Drug Law Institute described in the Introduction.

Nonetheless, concerns fostered by this event had limited effects. A quick glance at the U.S. agriculture sector today makes plain that anti-biotech activists in the United States did not have the same degree of impact that their European counterparts did across the Atlantic. Despite the burst of activism between 1998 and 2003, U.S. activists were ultimately incapable of stemming the tide toward GM crops or food, or reversing the U.S. government's pro-biotechnology policy orientation, or turning consumers away from GMOs in significant numbers. In 2006, almost 90 percent of the U.S. soy crop, 83 percent of the cotton crop, and 60 percent of the corn crop were genetically engineered, and thanks to the ubiquity of GM corn syrup, corn oil, and canola oil in processed food, GMOs had come to form part of almost every American's daily diet.[1] Indeed, if the U.S. anti-biotech movement is looked at from the vantage point of 2008, little evidence indicates that it had much of an influence at all.

Yet looks can be deceiving. In this chapter, we argue that the U.S. anti-biotech movement did have a significant effect on the agricultural biotechnology industry as well as on the trajectory of the technology, although not in ways that were as direct or obvious as the consequences of activists' actions in Europe. One realm in which the U.S. movement's influence was strongly felt was in the area of GMO commercialization. In the case of recombinant bovine growth hormone, for example, the movement put up such a fight that rBGH was kept off the market for several years. Activists were also instrumental in "killing" a GM potato by convincing the nation's fast food companies that this product wasn't worth the risk of an activist boycott, leaving the agricultural biotech industry without a customer for its potatoes. Perhaps most significantly, anti-biotech activism helped to keep Roundup Ready wheat, considered by Monsanto executives to be one of the company's best technologies, sitting on the shelf, even though it was ready to sell in 2002. In combination with the market shutdown in Europe, such actions not only imposed large economic costs on the industry but also altered its calculus about whether developing new GM products was worth the risk. Put in the terms used by the regulatory scientist quoted in chapter 2, activism raised the expected "expense-to-revenue ratios" for these companies' future projects, making them less attractive economically.

Interestingly, some of the U.S. movement's influence was felt most powerfully outside the country. U.S. activists played a major role in creating a global regulatory regime to govern trade in GMOs (the Cartagena Protocol on Biosafety) over the loud objections of the U.S. government (see chapter 6). These activists worked closely with others around the world, supporting their struggles against the technology. Finally, North American activists were among the key contributors to the critical analysis and alternative discourses upon which global opposition to agricultural GMOs was built (recall chapter 3). In short, while the effects of U.S. anti-biotechnology activism were admittedly more subtle, attenuated, and diffuse than those of their European counterparts, the U.S. movement *did* have a significant influence on the technology, both within the country and beyond.

In this chapter, we chronicle the U.S. controversy over agricultural biotechnology from 1975 to the present. We show how activism within the United States grew slowly but steadily from the late 1970s to 1998, when several events brought new resources and energy to the movement and catalyzed the major wave of activism with which we started this chapter. For the next five years after 1998, the U.S. anti-biotech movement pulsated with energy. It adopted new strategies and tactics and worked on multiple fronts simultaneously, contributing to the worried reaction the U.S. biotechnology industry expressed at the beginning of 2001. As activism gained momentum and foreign markets started to close, however, the industry reacted. Determined not to let what happened in Europe transpire in its own backyard, the U.S. biotechnology industry gathered its forces and organized against its domestic opponents. This coordinated reaction was an important means by which U.S. anti-biotech activism was undermined.

In addition to facing an aggressive response from the industry, the anti-biotech movement on U.S. turf was hampered by other features of the organizing environment. One was the strong support the government gave to the biotechnology industry. Most obviously, this support took the form of a conducive regulatory environment, but it also manifested itself in the executive branch's open enthusiasm for biotechnology. Another difficulty the activists faced was the close relationship that existed between the U.S.

biotechnology industry and other actors along the food commodity chain, particularly farmers and their commodity associations. This relationship was simultaneously social, economic, and cultural, and it reflected these actors' shared worldviews about the benefits of applied science and technology in the agricultural sector. A third major obstacle was U.S. consumer culture. Unlike European consumers, who were openly worried about the quality of their food and how it was produced, U.S. consumers showed more concern about the convenience and price of food than they did about whether or not it was genetically modified. This made consumer mobilization against GMOs impossible to catalyze. Hence, one of the tactics that had worked most effectively for European activists turned out to have no purchase in the U.S. context. Together, these phenomena worked to constrain the movement's domestic efficacy.

From Cambridge to California: Early Struggles against Biotechnology

As we demonstrated in chapter 3, the history of social resistance to genetic engineering in the United States is as old as the technology itself. The first struggle over biotechnology took place in Cambridge, Massachusetts, in the mid-1970s, around a new laboratory that Harvard University wanted to build for recombinant DNA research. Although the opponents of the lab failed to stop the facility from being built, their actions did bring the new technoscience of genetic engineering out of the realm of the scientific elite and into the public sphere. In the process of waging their highly public struggle against the Harvard lab, these academic scientists made it clear that biotechnology would not be apprehended as simply another step forward in the steady march of science. Rather, it would be an object of contention and social struggle.

The story of the Harvard rDNA laboratory controversy began in 1975, shortly after the Asilomar conference in Pacific Grove, California, where over one hundred of the world's leading biologists debated the opportunities and risks associated with rDNA research. In the wake of Cohen and Boyer's gene-splicing discovery, members of the U.S. scientific community agreed to place a de facto moratorium upon further rDNA research until its potential risks and dangers were more fully discussed.[2] Scientists worked out a set of recommended guidelines they believed would ensure the safe conduct of rDNA research. These guidelines were subsequently adopted by the National Institutes of Health (NIH), and the moratorium came to an end.[3]

By this time, many biological scientists were anxious to embark on their own gene-splicing research. Harvard sought to facilitate its scientists' work by planning to build them a moderate-risk-level research facility. But not everyone was persuaded that such research was safe to pursue. Among those who remained concerned were George Wald and Ruth Hubbard, two well-known biologists from the very same department that was pushing for the lab and some of their colleagues at the nearby Massachusetts Institute of Technology. Thus, in 1976 when Harvard moved forward with its plans, these scientists, together with concerned friends and colleagues from the organization Science for the People, sought to stop the lab from being built.

The first step they took was to organize an on-campus debate on the public health risks of genetic engineering. After failing to generate support for continuing the research moratorium, however, these activist–scientists decided to take the matter public.[4] They went to the Cambridge city mayor, Alfred Vellucci, to persuade him that the lab was a bad idea. Mayor Vellucci, who had never been keen on Harvard's elitist attitude anyway, put the issue on the agenda of the Cambridge City Council, which agreed to set up hearings.

During the hearings, the council imposed a moratorium on all moderate- to high-risk rDNA research, which temporarily tied Harvard's hands. But ultimately, the opponents of the facility failed in their efforts. According to Rae Goodell, who studied the controversy closely, Harvard and MIT administrators organized strategy sessions with scientists, public relations specialists, and lawyers to coordinate their efforts. They invited a number of national figures to the hearings "to impress the city audience on behalf of Harvard—'to sandbag them with more heavyweights,' as one proponent later bluntly put it."[5] Harvard and MIT also flooded the city councilors with calls and mobilized several Nobel Prize winners in support of the research. As a Harvard faculty member later reflected, "I remember being quite surprised at the vast resources they had to influence the community. . . . And I all of a sudden became aware of this whole organization that worked behind the scenes to influence the public."[6]

In the end, the Cambridge City Council Review Board recommended that the lab go forward. The board did require, however, that the lab's research be monitored. In 1977, the City of Cambridge codified these recommendations into the first rDNA law in the country. According to Sheldon Krimsky, the law "symbolized the right of local government to exercise control over where . . . research gets sited and the safety conditions of its performance."[7] But the real significance of the law and the controversy that gave rise to it was that it forced the issue out into the open.

While Harvard may have successfully defended its scientists' "rights" to unimpeded research, the critical scientific community who challenged the university's rDNA lab made sure they could not conduct their research without public knowledge and scrutiny.

The Struggle over a GM Bacterium

If the first concrete struggle over biotechnology in the United States was initiated by an elite scientific community, the second pair of struggles was initiated by a coalition of local and national activists with scientific expertise. Both of these challenges took place in California, and both involved the first GMOs to be field-tested in the environment: namely, a genetically engineered strain of a bacterium known as *Pseudomonas syringae*. The mutant bacterium came to be known as "ice-minus," to reflect that it was engineered to resist frost formation on plants.

In 1982, two University of California–Berkeley biologists, Steven Lindow and Nickolas Panopoulos, sought the approval of the National Institutes of Health to field-test the microbe. Their plan was to examine its efficacy by spraying it onto potato plants at a university farm in Tulelake, a tiny farming town near the Oregon border.[8] After a lengthy review process, NIH gave Lindow and Panopoulos the green light, and the two scientists prepared to carry out their experiment.

Closely following these developments was the Washington, D.C.–based anti-biotech activist Jeremy Rifkin. Immediately recognizing the significance of this federal approval for the first GMO field trials, Rifkin and his co-workers initiated a lawsuit against the NIH and the University of California in an effort to stop the trials.[9] The suit charged, first, that the University of California did not have enough insurance to cover liability costs for any damage that might occur and, second, that the NIH had failed to prepare the required environmental impact assessment. The federal district court judge agreed with their argument and in May 1984 imposed an injunction prohibiting the field trials from going forward. At that point, the Environmental Protection Agency (EPA) took over and began its own assessment of the experiment.

Over the next two years, local opposition increased markedly. Opponents formed a group called the Concerned Citizens of Tulelake (CCT) and pressured the Tulelake City Council and the Siskiyou County Board of Supervisors to prohibit the trials. In the midst of the turmoil, the EPA decided that the experiment did not pose an environmental threat and issued a field trial permit. The decision outraged many townspeople and prompted Rifkin's organization, the Foundation on Economic Trends, to file a new suit at the *state* level, seeking another injunction until California could conduct its own environmental assessment. Although the California judge granted a temporary injunction, the university researchers were ultimately given permission to go ahead with their test plots.

Some seven hundred miles south of Tulelake, local residents mounted a similar challenge against a biotechnology start-up by the name of Advanced Genetic Sciences (AGS). Shortly after Lindow and Panopoulos applied for federal approval to conduct their field test, this small Oakland-based firm submitted its own proposal to field-test a different strain of the ice-minus bacterium in Monterey County. In 1985, even before the EPA had approved the UC researchers' application, the agency notified AGS that it could proceed with field trials.[10] Again, Rifkin's legal team went on the offensive, requesting that the courts impose an injunction on AGS's experimental use permit. If the very first field trial permit could be cut off at the pass, reasoned the activists, that would make it easier to stop the biotechnology ball from rolling.

In March 1986, the judge in the case dismissed Rifkin's suit on the grounds that the plaintiffs were unlikely to convince the court that the EPA had violated the law. By this time, worried residents had expressed their concerns to the Monterey County Board of Supervisors, which banned all experiments within the county for a year. Advanced Genetic Sciences reapplied to the EPA in December 1986 to carry out its field trials in another county.[11] The EPA approved the application, and five months later AGS scientists donned head-to-toe space suits (per EPA requirements) and sprayed their ice-minus bacteria on strawberry plants near Brentwood, California. The press was on hand and brought their striking images to newspapers and magazines across the country.

Even though both the Berkeley researchers and AGS ultimately triumphed in these legal battles, the genetically engineered version of the microbe never made it to the marketplace. As Bernice Schacter later reflected,

> The results of tests over four years and ten frosts . . . indicated that indeed the ice minus bacteria would survive and would reduce ice damage by 80 percent. However, this success did not lead to the commercialization of genetically engineered ice minus bacteria . . . because of the significant negative publicity, the certainty that the approval process for commercialization would be long and expensive, and that Rifkin and the FoET [Foundation on Economic Trends] would continue to raise legal challenges to commercialization. In fact, AGS merged with another company . . . in 1989. [That company] decided that the legal challenges raised by Rifkin would be more

expensive than the sale of the [GE] bacteria would bring in profits and stopped research on ice minus in 1990.[12]

In essence, anti-GM activism had created so much controversy, regulatory hassle, and expense that it ultimately killed the project. The activists' ruckus also pushed the federal government to establish a formal regulatory policy for biotechnology. Biologist-created NIH guidelines may have worked for regulating physical safety in the lab, but the NIH was not equipped to evaluate or regulate GMOs once they moved beyond the confines of the laboratory. Nor did the Reagan administration want the relatively cautious EPA to be the deciding agency when it came to regulating this new technology, as we show below.

Establishing a Regulatory System

As the science of genetic engineering proceeded apace, the U.S. government, under Ronald Reagan, was forced to develop an official system of regulation. The Reagan administration was initially reluctant to regulate biotechnology, because the very idea of regulating industry ran squarely in the face of the administration's *de*regulatory bent and its corresponding political discourse. Since coming to office, the Reagan administration had advanced a powerful narrative about "freeing" the private sector from the red tape of regulation and letting U.S. companies get on with the business of making products and competing in the marketplace. A second source of reluctance stemmed from Reagan's desire to stimulate the economy. Economic growth had slowed during the 1980s, and American industry was losing its competitive edge. In this context, the Reagan administration did not want to enact measures that might dampen the growth of any new industry, especially one that was seen as fundamental to the country's economic future.

Two converging pressures finally pushed the Reagan administration to create a formal policy for regulating biotechnology. One was the momentum that had begun to build within the U.S. Environmental Protection Agency for a leading role in biotechnology regulation under the 1976 Toxic Substances Control Act. In the eyes of Reagan administration officials, EPA jurisdiction over biotechnology could be disastrous, because the agency's mission of environmental protection was likely to discourage investment and research in biotechnology and thwart the growth of this nascent industry. The Reagan administration was thus motivated to develop a biotechnology policy plan that would wrest as much control as possible away from the EPA and put it in the hands of the two other agencies that had legitimate claims for regulatory authority, namely, the Department of Agriculture and the Food and Drug Administration (FDA).[13] Both could be trusted to be far more sympathetic toward the new technology.

The other source of pressure came from certain companies within the biotechnology industry itself. Products were starting to come down the pipeline, and the companies producing them asked for government regulation.[14] In 1986, several officials from Monsanto paid a visit to the White House to meet with Vice-President George H. W. Bush to deliver the message personally that the industry wanted, indeed *needed,* to be regulated.[15] Not only would regulation protect companies against future liability claims, these industry officials believed, but also, and even more important, it would assure the public that this new technology was safe. Of course, Monsanto would be more than happy to lend its expertise and contribute suggestions about how that policy should be constructed.

The Reagan administration responded to these combined pressures by proposing the Coordinated Framework for the Regulation of Biotechnology. Coming out of the president's Office of Science and Technology Policy (OSTP), the Coordinated Framework suggested not that biotechnology needed to be strongly regulated but rather that it needed to be supported because of the significance of its economic and social potential. "The tremendous potential of biotechnology to contribute to the nation's economy in the near term, and to fulfill society's needs in the longer term, makes it imperative that progress in biotechnology be encouraged," noted the initial announcement in a 1984 *Federal Register.*[16] The administration indicated it was seeking to find a balance between protecting health and safety, on the one hand, and ensuring "regulatory flexibility to avoid impeding the growth of an infant industry," on the other.[17] In practice, the scales tilted much more heavily toward the latter.

Assuming from the outset that existing laws were adequate for regulating this new industry, the Coordinated Framework divided regulatory authority for biotechnology among the Food and Drug Administration, the Environmental Protection Agency, and the Department of Agriculture. In the process, it notably reduced the regulatory scope of the EPA, since only products that had the properties of a pesticide would now require EPA review, significantly diluting the influence of the only agency oriented toward serious precautionary regulation (the EPA), and strengthening the hand of the two agencies that were primarily concerned with promoting U.S. agroindustrial development (the USDA and FDA). The importance of this restriction on the EPA's power became manifest in the final version of the Coordinated Framework, which crucially mandated that the biotechnology industry be regulated on a product-by-product basis and not on the basis of the *process* by which these products were made (i.e., through the techniques of genetic engineering).[18] Underlying this policy approach was the presumption that genetically engineered organisms were "substantially equivalent" to organisms produced through traditional breeding methods and did not require any additional oversight or regulation. For activists, then, the framework made it impossible to challenge GMOs as a *class* of products and to establish precedents that covered all products produced with modern methods of biotechnology. Every future fight over a GMO would thus have to be fought de novo.

After 1990, the incoming administration of President George H. W. Bush further relaxed the nation's regulatory controls on biotechnology. This Bush administration strengthened the language of "substantial equivalence" of GM and non-GM products and increased the burden of proof on those worried about the health and environmental impacts of GMOs. After 1992, regulatory oversight would be exercised only "if the risk posed by the introduction [of a GMO] is unreasonable, that is, when the value of the reduction in risk obtained by additional oversight is greater than the cost thereby imposed."[19] The regulatory retreat was particularly apparent in the FDA's new policy for approving genetically modified foods. Under the new policy, most foods from GM plants would be "generally regarded as safe" and would not require any special approval before being allowed on the market; all a company need do was to meet with the FDA to discuss the safety and nutritional aspects of its product.[20] A year later the USDA introduced a simplified notification process for field trials of some GM plants, facilitating their testing. The slant of regulatory policy was clear. Most companies were sufficiently foresighted (and risk adverse) to provide an immense portfolio of data to the relevant federal regulatory agency before attempting to go to market, but regardless of this substantiation, they could rest assured that these agencies were not in the business of trying to block their products.[21] If anything, both groups shared a similar goal: to get the data in hand to facilitate the biotechnology industry's success. The number of crops receiving government approval in the 1990s (some fifty-three by 1999) was a clear indication of the government's pro-biotechnology stance.[22]

In short, every administration from Ronald Reagan's through Bill Clinton's viewed biotechnology in terms of its potential to carry the U.S. economy into the twenty-first century. In the best-case scenario, the biotechnology industry would help the country regain its international competitiveness and reanoint the United States as a world leader in technology development. While each administration knew that its job included the oversight of this new and rapidly advancing technology, not one had any interest in slowing the biotechnology train down. With a government so positively disposed toward biotechnology, the anti-biotech movement faced a major obstacle when it came to challenging the deployment of GMOs in the United States. Of course, that did not stop activists from attempting to use the official regulatory system as a means to constrain the technology. The struggle over bovine growth hormone offers a case in point.

The Struggle over rBGH

The struggle over recombinant bovine growth hormone, or rBGH, began in 1986, just as the ice-minus battle was reaching its peak. Although dairy scientists had long known they could increase milk production by giving cows an added dose of a hormone that they already produced in their own pituitary glands, they also knew that the natural sources of this growth hormone were too limited to be commercially viable. The advent of genetic engineering relaxed that limitation, however. In the early 1980s, research scientists at four large chemical and drug companies—Monsanto, Eli Lilly, American Cyanamid, and Upjohn—all embarked upon major research projects to produce bovine growth hormone synthetically, collectively spending several hundred million dollars on the effort.[23] At stake was a market estimated to be worth from five hundred million to one billion dollars a year.[24]

Monsanto made it to the finish line first and promptly applied to the U.S. Food and Drug Administration for permission to market its new product under the trade name Prosilac. Its competitors soon followed with their own versions of the synthetic hormone. Although the FDA was not yet ready to approve any of these versions of rBGH for commercial use, its initial reaction to the hormone revealed the administration's interest in supporting the country's fledgling biotechnology industry. After reviewing the available evidence from field trials, much of which was still preliminary and had come from the companies developing rBGH or from those whom it had contracted to test the product, the FDA concluded that the synthetic hormone posed no substantial health risk to human beings or to animals.[25] In 1985, it thus allowed rBGH field-trial milk to be mixed with non-BGH milk and sold to the public without consumers' knowledge.[26]

Two groups had been following the development of rBGH closely. One was Jeremy Rifkin and his fellow activists at the Foundation on Economic Trends. The FDA's approval of rBGH would make it the first new biotechnology product available on the market. Thus, for these anti-biotechnology activists, rBGH represented a crucial line in the sand, or as Rifkin put it, a potential "floodgate" through which other biotechnology products would flow.[27] Given the significance of rBGH's approval, Rifkin decided to throw all of his organization's resources into fighting it.

Around the same time, a number of small-scale dairy farmers and their supporters from traditional dairy states in the Northeast and Midwest started mobilizing. The small-scale dairy states of Vermont, Wisconsin, and Minnesota were already feeling the pain of increased output as large dairies in the South and West entered the industry and helped to flood the country with milk. Indeed, the very same year Monsanto began seeking FDA approval for rBGH, the USDA initiated a $1.8 billion program to pay participating dairy farmers to slaughter over a million cows to relieve financial pressure on the industry.[28] From these dairy farmers' perspectives, introducing a technology that would *increase* milk output would only exacerbate their problems by driving milk prices down even further. As Wisconsin dairy farmer Mike Cannell aptly put it, "Are we ready for 5,000 dairies with 1,500 cows each producing all the milk we need? . . . This whole fight is about what we want American agriculture to be and what we want rural America to look like."[29] In 1986, Cannell and other dairy farmers formed a coalition with Rifkin so that they could fight rBGH together.

As usual, Rifkin started off with a procedural attack on U.S. government policy. As soon as Monsanto and its competitors sought FDA approval to market rBGH, Rifkin filed a petition with the FDA charging that farmer adoption of the growth hormone would "damage the environment, cause unnecessary suffering to cows, and wreak havoc on the dairy community."[30] Rifkin demanded that the FDA prepare an environmental impact statement before it approved these requests. Signing the petition were the Wisconsin Family Farm Defense Fund, the Humane Society of the United States, and Wisconsin's secretary of state Doug LaFollette.

Over the next six years, Rifkin's group undertook a number of other rBGH-related actions against the government, including a 1990 suit filed against the USDA for using public monies to support a million-dollar public relations campaign promoting the growth hormone.[31] Rifkin also focused on the consumer end of the commodity chain, circulating a letter to a dozen of the nation's top supermarket chains, asking them to clarify their policies toward rBGH milk for the public record.[32] In addition to noting that rBGH milk from test cows was coming into the milk supply without consumers' knowledge, Rifkin informed these supermarket chains of a yet-to-be-published paper by a University of Chicago physician and professor of medicine that exposed the human health risks of the synthetic hormone.[33] Anxious about the consumer reaction, spokespeople for Safeway, Kroger, Stop and Shop, Pathmark, Supermarkets General, Vons, and several major dairy product producers (Kraft, Borden, and Dannon) publicly declared that their companies would not sell or make dairy products treated with the hormone.[34]

At the local level, small-scale dairy farmers who were opposed to the technology teamed up with family farm defenders, a handful of concerned legislators, and some "alternative" food companies to fight rBGH. In Vermont, Ben and Jerry's took a proactive stance against the growth hormone, agreeing to pay its suppliers a premium for non-rBGH milk and criticizing rBGH's likely impact on the state's small dairies. The Vermont Legislature also passed a mandatory rBGH labeling law (which was later declared illegal).[35] In Minnesota, food activists successfully moved a bill through the legislature that placed a ban on the use of rBGH for a year.[36] But it was in Wisconsin that the battle over rBGH was the fiercest and longest lasting. There, a coalition of mainly grassroots groups led by a small dairy farmer named John Kinsman vigorously challenged rBGH for six years. In the process, they raised many consumers' awareness of the technology through the state and national press coverage they attracted.

Industry Reactions

Initially, the industry was caught off guard by the activists' attacks on rBGH. But it quickly recovered, developing a multipronged strategy to squelch the opposition and defend its right to make and market its product. From the outset, the industry and its lobbying organization, the Animal Health Institute, sought to control the discourse around rBGH. Rather than using the alarming term *recombinant bovine growth hormone,* for example, the industry began using (and urged others to use) the more technical term *bovine somatotropin* in its stead. In addition, the industry attempted to convince respected opinion-makers, such as physicians and other health professionals, that rBGH had no negative impacts on human or animal health. In 1988, for example, the Animal Health Institute distributed sixteen thousand fliers on the safety of rBGH to doctors and other health professionals with the idea that these recipients would be seen as impartial and legitimate experts.[37] They also mailed out seventy-two hundred copies of a sixteen-page brochure.[38] As James Brezovec, the marketing director for Monsanto's dairy products, explained, "The marketing strategy is to get the facts to those whom the consumer will look to for advice."[39]

A second component of the industry's strategy involved maintaining close contact with dairy processors and milk marketing associations, relying on the business and social relationships with the people in these organizations to gain their support. Because both groups shared a basic belief in the benefits of new technology and the market as the appropriate arbiter of a technology's utility, it was not hard to convince them that rBGH should be made available to dairy farmers and its fate should be decided in the marketplace rather than in the halls of government. Indeed, for most participants in the dairy industry, supporting rBGH was a no-brainer: it had proved itself efficient in terms of the nutrients required to produce a given quantity of milk, the milk it yielded was identical to non-rBGH milk, and it represented the cutting edge in dairy science. In the words of one dairy economist, virtually the whole industry allied itself behind the technology for the simple reason that "the industry has been drilled with [the idea of] efficiency. . . . Science is very important in this industry. Producers, processors, co-ops, dairy farmers . . . the whole sector, up to the retail sector, was supportive."[40]

Finally, when legislation to ban the use or sale of rBGH or to require labeling was brought up before various state legislatures, the industry and its lobbyists lobbied vigorously against the bills, talking to key legislators, meeting with agricultural committee members, and working with state dairy associations to defeat these efforts. In an interview with a *St. Louis Post-Dispatch* reporter, Wisconsin state senator Russell Feingold, who sponsored several bills against rBGH, noted that the lobbying by Monsanto and others against the labeling bill "was the most sickening display of corporate aggressiveness in the history of Wisconsin."[41] According to the reporter who penned the story, Feingold was referring to the "swarm of lobbyists hired by the chemical companies to put pressure on Wisconsin dairy distributors and factories."[42] Monsanto used economic threats to create allies in the struggle, telling universities they would lose research money, telling distributors that separating rBGH from non-rBGH milk would be prohibitively costly, and telling dairy producers and plant workers that they would lose business and jobs if any labeling laws passed.[43] Labeling was seen as the kiss of death by the industry, because it enabled consumers to identify GM foods and vote with their dollars if they decided they did not want them. It thus had to be stopped before it gained traction.[44]

In the end, despite the immense fight the anti-biotech activists waged and the problems they caused for the industry, the companies proved far more powerful. Through political pressure and their successful efforts to create allies, the biotech companies managed to defeat nearly every bill designed to restrict rBGH's use or sale or to make it identifiable by consumers through labeling. The companies also had a powerful impact on the discourse around the technology and the way it was presented to the U.S. public by government and the media. In November 1993, the FDA finally approved rBGH, and in February of the following year, Monsanto started selling Prosilac to U.S. dairy farmers. Not all farmers bought in to the technology, but enough did to secure rBGH's profitability for the companies.[45] Over the following decade, the debate died down, and rBGH no longer made headlines. Processors and supermarkets gradually reverted to a "don't ask, don't tell" policy toward the growth hormone, quietly backing away from earlier claims that they would not manufacture or sell dairy products made from treated cows. Consumers also revealed their complacency toward rBGH, and in most parts of the country they purchased and consumed milk without regard to its method of production. Not until a decade or so later did the issue arise again.

Yet perhaps even more important than the industry's apparent "win" on rBGH were the lessons different groups drew from the struggle. In the biotechnology industry's reading, rBGH became an issue in the late 1980s not because the American public was afraid to consume food products made with the aid of new technologies but because a small group of environmental and animal rights extremists and a few disgruntled dairy farmers had made it into one. Indeed, the lack of a strong consumer outcry about rBGH suggested to Monsanto and other companies that American consumers were not likely to rebel against the use of genetic engineering in agriculture. Put quite simply, if they did not react negatively to a product like milk, which American parents obsessively feed to their kids, they were unlikely to lose sleep over GM corn or soy. That said, the industry did take away another lesson from this and the other early biotech struggles, which had to do with the character of the opposition. People like Jeremy Rifkin and John Kinsman may have been few in number, but their commitment to the cause, their determination, and their strategic skills made up for the thinness of their ranks. The industry was rightly put on guard and quickly came to see the importance of developing its own united front and strong public relations machinery to defend its interests.

For their part, the anti-biotech activists learned much about the power and modus operandi of their adversaries. They discovered the enormous lengths to which these large corporations were willing to go to protect their economic interests and the depth of support they had within the U.S. government and agricultural sector. They also learned that without strong evidence of a public health threat, it was hard to mobilize Americans against the technology. Arguments about the adverse impacts of the new biotechnologies on small farmers were not going to be enough to elicit a public reaction. Nor could the activist community count on the FDA, the USDA, or the EPA to factor in structure-of-agriculture issues or, for that matter, consumer right-to-know or animal welfare concerns when it came to evaluating the technology. The government had chosen to take a strictly "science-based" approach to biotechnology, which by definition pushed all *but* human health concerns off the table. In the process, it made itself immune to the activists' most profound critique of biotechnology, which involved its social and economic consequences and the undemocratic nature of decision making around it.

In sum, the battle over rBGH revealed, better than any other, the immensity of the challenge the activists faced as they sought to stop the development and deployment of biotechnology in the United States. Not only were they confronting a deep-pocketed, politically powerful, and committed industry that had invested a huge amount of money in this new technology and had every intention of defending it, but they also faced a government that had its own interests in seeing the technology and industry succeed. As if this were not enough, many of the nation's farmers proved to be highly supportive of biotechnology as well.

Enrolling Farmers

By the time rBGH received FDA market approval in late 1993, the industry had turned its attention to developing its next set of biotechnology applications. Among them were what would turn out to be the blockbuster products of the 1990s: namely, crops that were genetically modified to tolerate the application of specific herbicides, such as Monsanto's Roundup Ready and AgrEvo's Liberty Link corn and soy and crops that had been engineered to contain a natural pesticide known as *Bacillus thuringiensis (Bt),* such as *Bt* corn and *Bt* cotton. By choosing to invest in these particular applications, Monsanto, for one, sought to meet its own criteria for making large-scale R&D investments, as described by the former official we quoted in chapter 2: these were products that solved real problems for farmers, they could potentially supply enormous markets, and they could generate demand over the long term. The only remaining question was whether these products would meet the fourth criteria: Would farmers perceive them as being worth the added expense and be willing to pay for them?

The answer, it turned out, was a definitive yes. Much to the activists' chagrin, when these new genetically engineered crops came on the market in the mid-1990s, U.S. farmers practically lined up to buy them. Farmers adopted these GM crop varieties faster than they had adopted any other agricultural technology in the nation's history.

American farmers' enthusiasm for GM crops could be traced to several sources. First, these technologies had been developed with American farmers and their specific farming challenges in mind and offered many farmers what they perceived as real advantages. According to a survey done at the time, many adopters felt that the new GM crop varieties would raise yields by reducing pest problems and lowering pesticide costs.[46] In the case of *Bt* corn and cotton, the new technologies were also labor-saving. Rather than having to spray their fields with pesticides multiple times each season, they could cut their pesticide applications significantly.

Aside from these material attractions, many farmers were attracted to the new technologies for cultural reasons. Like most farmers around the world, U.S. farmers took pride in the appearance of their fields. By killing everything in the field but the genetically engineered plant, herbicide-tolerant crops helped produce neat and weedless fields, in spades. Biotech crops also spoke to U.S. farmers' welcoming attitudes toward science and technology. Here was a new technology that promised to increase yields, lower costs, and make their crops more appealing. Not only that, it reflected the latest science and was offered by companies whose products they had been using for decades.

Even with U.S. farmers already predisposed to accept the technology, the agricultural biotechnology companies were not about to leave the success of their new crop technologies up to chance. Prior to rolling out its GM seeds commercially, Monsanto made a strategic decision to cut seed dealers and distributors in on the financial benefits of its new, genetically modified Roundup Ready corn and soy products.[47] It worked closely with agro-input distributors as well as middle-order seed producers who used its licensed technology, inviting them to sales-related luncheons and offering them economic incentives for selling more GM seeds. Monsanto also maintained direct access to farmers through dense networks of company representatives who would regularly visit midwestern farms. It carried out extensive field trials with farmer participation, showing farmers firsthand how these technologies worked to address problems such as the corn borer, a serious pest in the Midwest. According to one salesman who worked for an agricultural supply company, Monsanto did a remarkable job in developing close relationships with the farm community through its representatives, who were knowledgeable, well trained, and often well liked. The company was viewed as competent, reliable, and solid among farmers, even while seed companies resented Monsanto's efforts to limit their own profits through a competitive distribution system.[48] In effect, the company built strong relationships with farmers who believed in the quality of its products and were prepared to pay a premium price for them.

The Struggle Heats Up (1998–2003)

Undaunted by their loss of the rBGH battle and by midwestern farmers' positive reactions to the technology, U.S. anti-biotechnology activists continued their struggle against GMOs. Toward the end of the 1990s, the movement's strength, energy, and momentum reached a crescendo. This upsurge in anti-biotech activism reflected a combination of phenomena. As more activists working in the critically minded NGO community learned about genetic engineering, many convinced their organizations to let them devote time to the issue. This brought new activists and organizations into the movement. The optimism generated by the successes of the European anti-biotech movement also served as a powerful motivator. When U.S. anti-biotech activists opened e-mails from their European counterparts, read the news from Europe, and talked to their counterparts about what they had achieved there, it gave them a renewed sense of hope and energy for their own struggle.

A final factor undergirding the boom in activism was an increase in funding. In 1997, a longtime foundation director who was highly critical of agricultural biotechnology was asked to help coordinate the Association of Environmental Grantmakers' meeting to be held in Houston, Texas, that year. Sensing a unique political opportunity, he invited several university scientists who were critics of genetic engineering to speak to funders at the meeting about the environmental significance of the new genetics. About twenty-five environmental grantmakers attended the talks and were so swayed by what they heard that they decided to establish the Funders' Working Group on Biotechnology. Over the next three years, both the working group and the individual foundations that supported it poured an estimated two million to three million dollars into the cause.[49] This added to the movement's momentum and enabled it to engage in a broader range of contentious actions. It also spawned two important new anti-GM coalitions: the large grassroots-based Genetic Engineering Action Network, or GEAN, and the more elite and centrally coordinated Genetic Engineering Food Alert, or GEFA.[50] Both helped to bring anti-biotech activist groups together and facilitated their coordination.

With new resources and energy flowing into the movement, activists expanded both the number of their attacks and their tactical repertoire. One tactic the movement pioneered was the testing of products for the presence of GMOs. In the fall of 2000, Friends of the Earth sent several major brands of taco shells to be tested for the presence of Starlink corn, which had been approved only for animal consumption, not human. When the tests came back positive, the activists went directly to the press, where their discovery set off a huge national and international reaction (which we discuss further below).

Greenpeace USA and GEFA also spearheaded a new series of campaigns aimed at pressuring U.S. food manufacturers and retailers to re-nounce the use of GM products. These groups chose several major food manufacturers (e.g., Gerber, Kraft, and Kellogg) and urged consumers to boycott their products until these companies agreed to stop using GMOs in their breakfast cereals and other processed foods. They also launched supermarket campaigns aimed at Safeway, Shaw's, and Trader Joe's in the hope that they could push them to go GMO-free. Now that GMOs were circulating more widely through the U.S. food supply and the news of the European consumer rejection was spreading, activists hoped to trigger a similar consumer response in North America.

Serendipity helped to generate a growing sense of momentum within the movement. In May 1999, a Cornell University entomologist by the name of John Losey and his colleagues published an article in *Nature* illustrating that the pollen produced by *Bt* corn killed a significant pro-portion of monarch butterfly larvae in a laboratory experiment.[51] The Losey study caused a worldwide stir. People all over the world loved and extolled the monarch and thought of it as the "Bambi" of the insect world, a species that deserved protection. From the movement's perspec-tive, Losey's finding was akin to a gift from heaven. As one anti-biotech activist remarked, "A colleague called me up and said, 'You are not going to believe this one. They just found out that *Bt* corn pollen kills mon-archs!' And we said . . . it's horrible that it's going to kill an insect, but it's a charismatic insect, and it's an insect that everybody learns about. It [was] the best news we had heard in a long time."[52] The repercussions were enormous. For years, the story circulated in the news, and activists used the monarch symbol at virtually every anti-biotechnology demon-stration. And even though the industry harshly attacked Losey's research methods, the story took on a life of its own and one that highlighted a key concern of the movement: the potential environmental threats associ-ated with GMOs.[53] As the activist just quoted explained, "It didn't matter whether or not the science was good or bad; it just gave people and the press a glimpse of the possible seriousness of the impact. For the first time, [there] was something that people could grab on to, and relate to, on a personal level. . . . For the first time the threats were made real."

Movement Gains and Losses

Thanks to its full-court press and incidents such as the Losey study, the movement had scored some significant gains against the biotechnology industry. On July 30, 1999, for example, the *Wall Street Journal* published a front-page article featuring Greenpeace USA's concerns that Gerber baby food contained GM ingredients even though its Swiss parent company at the time, Novartis, had removed GMOs from the products it sold in Europe. Worried that the issue would explode in its face, Novartis quickly declared it would stop using genetically engineered ingredients in its Gerber line.[54] Around the same time, McDonald's Corporation and sev-eral other fast food chains, worried about consumer acceptance of GMOs, responded to a threatened activist boycott by informing their suppliers that they would not purchase GM potatoes for use in their french fries. As a result of these companies' actions, Monsanto decided to take its new GM potato (called the NewLeaf) off the market and close down its GM potato research plant in Bangor, Maine. In January 2000, the giant snack maker Frito-Lay become so worried about the market uncertainty caused by activist pressure that it formally declined to buy genetically altered corn for its snack foods.[55]

Coming on the heels of these events, the discovery of the unapproved Starlink corn in the country's taco supply wreaked even greater havoc on the agricultural biotechnology industry. As a result of the Starlink exposé, food manufacturers were forced to recall nearly three hundred products, U.S. corn trade with many foreign countries was seriously disrupted, and Aventis Crop Science—the company responsible for the leak into the human food supply—was subjected to numerous lawsuits for its negli-gence. Some five months after the corn was detected, the USDA agreed to spend up to twenty million dollars to buy back the remaining Starlink corn seed from farmers and seed companies. Despite the efforts of the industry and the Bush administration to resolve the crisis, reverberations from Starlink were felt long after the problem was identified. In the end, the debacle cost the industry and the U.S. government millions of dollars and threw the safety of U.S. agricultural exports into question.[56]

One arena in which the movement's success was distinctly limited, however, was in its campaigns against U.S.-based supermarkets and food manufacturers. Although the movement did manage to induce the California-based supermarket chain Trader Joe's to go GM free, all the rest of its consumer campaigns failed. Pressure on Kellogg to reject GM corn in its breakfast cereals was simply shrugged off. Kraft made no move to change its policies when it was targeted. And despite efforts targeting the Safeway and Shaw's supermarket chains, the movement never managed to convince either company to reject GM foods.

Activist campaigns aimed at U.S. supermarket chains and food manufacturers were difficult to carry out successfully for several reasons. The most important was a U.S. food culture that made it hard to mobilize American consumers—whose preferences strongly influenced the decisions of supermarket managers—against genetically modified foods. Until recently, U.S. food culture has historically been shaped by a preoccupation with price and convenience, rather than a concern with the quality of the food, including whether or not it is produced with GMOs. In general, U.S. consumers knew little about GM foods and, in the absence of any food catastrophe or regulatory disasters, had no obvious cause for concern. Thus, when anti-biotech activists sought to convince U.S. consumers that GM foods were unsafe and untested, they could not capture the public's imagination. Few consumers contacted their supermarket managers and food manufacturers to express their distaste for this new agricultural technology.

The relatively high level of trust that many Americans place in their food regulatory authorities contributed to consumer indifference. Indeed, a 2001 nationwide survey funded by the Pew Foundation revealed that 41 percent of American consumers trusted the U.S. Food and Drug Administration "a great deal" when it came to information about GM foods, and another 44 percent trusted it "some." Only 8 percent reported that they did not trust the FDA "at all."[57] In the context of such trust, activists had a tough time convincing the public that GMOs were dangerous enough that they should be kept out of consumers' food pantries.

The movement's inability to succeed with its consumer campaigns also reflected the structural and cultural characteristics of the supermarket sector. Because the U.S. supermarket sector comprised some 24,600 stores spread out over a half dozen regional markets, activists found it impossible to target the sector comprehensively. Moreover, in contrast to most European supermarkets, the U.S. supermarket industry had taken a "low-price" rather than "high-quality" strategy when it came to establishing their private brands. American store brand labels thus attracted customers whose primary concern was the products' prices rather than their special qualities. This rendered the U.S. supermarket industry invulnerable to the kind of tactics that had worked so well in Europe, in which activists helped to create a strong association between high-quality supermarket brands and products that were GMO free.

A final obstacle lay in the cultural predisposition of both the U.S. supermarket and food manufacturing industries to accept GM foods. Such a predisposition arose from the close working relationships that exist among participants in the food chain and from the shared view that scientific innovations in the food sector are good for producers as well as for consumers. As an official from the Grocery Manufacturers Association (a major industry lobbying group) argued before the California Senate Agriculture Committee on March 28, 2000:

> The U.S. food industry enjoys an enviable record of producing and distributing . . . an abundant and affordable supply of food that is safe, wholesome, and nutritious. This goal has been achieved not only because of our fundamental commitment to food safety . . . but also through advancements in science and technology. By combining these discoveries with safety assurance and quality control techniques, U.S. food companies continue to produce new, better and safer foods. *GMA believes that modern biotechnology is the most recent in a long line of such advancements that will enable the food industry to produce and market these foods to consumers in America and around the world.*[58]

Because the supermarkets and food manufacturers shared a similar worldview, it was natural that they readily got onboard with biotechnology.

Fighting Back, with Verve

With Europe's market closure, growing anti-GM rumblings in Asia, and stepped-up activism in the United States, the agricultural biotechnology industry found itself in a very stressful situation at the turn of the century. Even though profits from almost two decades of investment in genetic engineering had finally materialized, the industry was coming under attack from all sides. "The protest industry has gone too far," Edward Shonsey, chief executive at Novartis Seed, told a *New York Times* reporter in November 1999. "They've crossed the boundaries of reasonableness, and now it's up to us to protect and defend biotechnology."[59] In this context, seven of the world's largest agricultural biotechnology companies agreed to put competition aside and to work together to ensure the survival of the industry.[60] They established an organization called the Council for Biotechnology Information (CBI), whose objective was to counter the opposition and to restore confidence in the technology. They committed a total of thirty million to fifty million dollars a year to the organization and hired away Cargill's vice-president of public affairs, Linda Thrane, to be its executive director. With twenty-five years of experience in strategic communications, Thrane was well-qualified to convince the American public of the benefits of agricultural biotechnology.

The new Council for Biotechnology Information launched a major public relations campaign to do precisely what Edward Shonsey had declared was necessary: protect and defend biotechnology. The CBI hired a global company, BSMG Worldwide, to be the group's advertising agency and placed a series of ads promoting biotechnology on the major television networks and in the print media. The CBI also conducted tracking surveys of the public and opinion leaders, seeking to find out what the American public "knew, and needed to know," about agricultural biotechnology.[61] The organization's new Web site, whybiotech.com, provided fact sheets and other information that could be easily accessed by the press and other interested parties. In an interview, Thrane explained her plan for marketing the benefits of biotechnology: to try to "reach the gatekeeper," namely, "the woman with two to three kids who basically controls the family's food-purchasing dollar," and to "get good information to people who haven't made up their minds so they come to their own conclusion that biotechnology offers them real benefits."[62] The organization was housed in Washington in the same building that housed the Biotechnology Industry Organization (BIO), a politically powerful lobbying association to which these same seven companies also belonged.[63]

The industry also fought back in other ways. From the late 1980s on, the U.S. biotechnology industry and its allies in the food manufacturing sector had lobbied hard, and successfully, for a regulatory approach to GMOs that emphasized the similarities (or "substantial equivalence") between GM and non-GM products rather than the distinctions between them. Such an approach, they argued, made labeling unnecessary and prohibitively expensive, a position with which the U.S. government concurred. This strategy was complicated in 2002 when anti-biotech activists tried an end run around this federal labeling policy, focusing on *state* labeling requirements instead. That year, a group of Oregon activists put a measure on the ballet (Measure 27) that would require mandatory labeling of all genetically modified foods sold in the state, insisting that Oregonians should have the right to know what they are eating.

The industry responded swiftly. Acutely aware that labeling would enable consumers to identify and avoid its products, the industry formed a corporate coalition comprising Monsanto, DuPont, General Mills, H. J. Heinz, and other food companies to fight the measure. The coalition poured $5.5 million into an intensive advertising and lobbying campaign. As part of the campaign, the coalition broadcast TV ads in the weeks leading up to the election featuring a grocer drowning in red tape, a farmer fearful about the future, and a doctor assuring viewers that genetically engineered foods are safe.[64] It also mobilized the support of its allies in the Oregon government and the FDA, which warned that labeling was illegal, because it would violate interstate commerce rules. Thanks to this concerted effort, the industry managed to protect its interests from another assault by the activists. The ballot was defeated by a large margin.

The Battle over Genetically Modified Wheat

It was in the midst of these struggles that the U.S. anti-biotechnology movement, in conjunction with a group of midwestern farmers, mounted another challenge to the industry. This challenge aimed to stop the deployment of a hard red spring wheat that Monsanto had engineered to tolerate the company's best-selling herbicide, Roundup. Planted on sixty-three million acres of U.S. farmland during the 1999–2000 growing season and representing the country's third largest commodity crop, Roundup Ready wheat promised to be the industry's next big blockbuster product, especially if U.S. farmers turned out to be as excited about it as they were about Roundup Ready soy and canola.[65]

Monsanto officials got their first strong hint that there might be a problem with deploying GM wheat in 2000.[66] In that year, two North Dakota legislators, at the behest of a coalition of concerned wheat farmers and anti-GM activists, introduced a bill that would impose a two-year moratorium on the commercial introduction of GM wheat in the state. According to Representative Phillip Mueller, one of the backers of House bill 1338, it was simply too risky for a wheat-dependent state like North Dakota to plant GM wheat when compelling evidence indicated that world markets would not accept it. If Starlink had taught the farm community anything, it was that moving ahead of the market was dangerous. "This issue is very simple, our potential market loss," Representative Mueller noted. "We don't really need any other excuses for our markets to get lower than they are."[67] Most of the House of Representatives agreed. With surprisingly little fanfare, HB 1338 passed the House by a vote of sixty-eight to twenty-nine and then moved on to the Senate.[68]

By the time the bill reached the Senate, word of HB 1338's passage in the North Dakota House had traveled east of the Great Plains states to create a stir in St. Louis, the home of Monsanto headquarters. Realizing that the North Dakota legislation could throw a monkey wrench into its business plan, the company lost no time in dispatching some officials to North Dakota to convince the state's senators and citizenry that passing such a bill would be a huge mistake and could cost the state the company's wheat research support.[69] "If this legislation passes," proposed Michael Doane, industry affairs manager for Monsanto, at a packed meeting at the Capitol, "I simply cannot ask [Monsanto] to continue to fund . . . the research that is necessary to develop biotech wheat in North Dakota." With the help of the powerful state Senate Agricultural Committee chair Terry Wanzek, the company managed to get the bill watered down to a nonbinding interim study resolution.[70]

The original bill for a moratorium may have been killed by Monsanto and its allies in the Senate, but farmer resistance to the introduction of GM wheat certainly was not. Although HB 1338 was formally introduced into the legislature by two sympathetic representatives, its main support actually came from a group of North Dakota and Montana wheat farmers worried about the threat GM wheat would pose to their viability. Organic wheat farmers were most concerned, knowing that if their seed stocks were contaminated by GMOs, their crop would be rejected by their buyers, and their organic premiums, market, and livelihoods would all evaporate in an instant.[71] With the help of the Dakota Resource Council, the Western Organization of Resource Councils (WORC), and a few local sustainable-agriculture groups, these farmers started to organize at the local, state, and regional levels.[72] In 2001, WORC held a summit meeting at its office in Billings, Montana, to bring these groups together with some national anti-biotech groups and Canadian activists who were also working on the GM wheat issue. Together, they mapped out a strategy for challenging the biotechnology industry giant. Local activists would do grassroots organizing, the national groups would catalyze broader support, and the Canadian activists would carry out a parallel challenge to Monsanto's effort to introduce GM wheat into Canada. For the next four years, these groups collaborated, with the local organizations taking the lead.

The organizing strategy these groups put together had several complementary components. One was to make clear to farmers the threat that GM wheat posed to their export markets and that they, rather than Monsanto, would be the ones left holding the bag (or more to the point, the bushel) in the likely event that major U.S. wheat importers, such as Japan and Europe, rejected GM wheat. To strengthen their claim, these activist groups distributed information that a wheat marketing group had collected about different countries' resoundingly negative reactions to GM wheat.[73] WORC also commissioned an Iowa State University economics professor, Robert Wisner, to study the impact that Roundup Ready wheat would have on wheat exports and prices. Wisner's study indicated that wheat farmers would experience a major foreign market loss and a decline in prices if GM wheat was introduced into the United States within the following two to six years.[74] WORC issued a press release with these findings and used it as an important farmer organizing tool.

The Center for Food Safety, one of the national groups involved in the anti-GM wheat coalition, assumed the task of petitioning the USDA to carry out an environmental impact statement before it deregulated Roundup Ready wheat. The petition demanded that the agency identify the "potential socioeconomic, agronomic, and environmental impacts that would result if the agency were to approve Monsanto's application . . . and develop means of mitigating the adverse effects of genetically engineered wheat."[75] The CFS included the Wisner report as supporting evidence for its petition and brought in several North Dakota and Montana wheat farmers as plaintiffs.

The activist groups from the Great Plains area also kept up their legislative efforts. In 2003, they got bills introduced into the Montana, North Dakota, Kansas, and South Dakota legislatures that would have thrown a roadblock in the way of any company trying to introduce GM wheat into these states. For example, HB 409, the Montana Wheat Protection and Promotion Act, would have required certification by the Montana Department of Agriculture before GM wheat could be introduced. For GM wheat to be certified, a company had to demonstrate that the state would experience clear net economic benefits from its introduction, that the company could cover any financial liability associated with a market or environmental problem, and that a system was in place to ensure public participation in decision making. Although the bills in each of these states were ultimately squashed, their very existence added to the groundswell of resistance to GM wheat. So did the organizing efforts of the Canadian activists, who had put together a large and diverse coalition of groups opposed to the introduction of GM wheat in Canada.[76]

At first, Monsanto responded to these challenges by trying to bring as many related groups around to its position as possible. In an effort to drum up support, it organized meetings with its U.S. allies, including the National Association of Wheat Growers, legislators, and producer groups; set up an advisory panel with farm groups in Canada; and made public presentations on the benefits of GM wheat.[77] It also signed a pact with Spring Wheat Bakers, a farmer cooperative owned by twenty-eight hundred members, to create an "identity preservation" system that could, in theory, separate GM and non-GM crops for the market.[78] In early 2003, in an effort to calm the agricultural sector's fears, the company publicly stated that it did not intend to release Roundup Ready wheat until consumers, growers, and buyers had no serious objections.[79] On May 10, 2004, however, Monsanto abruptly shifted gears and announced that it was deferring its plans to offer genetically modified wheat to farmers. According to the company's press release, it was "realigning research and development investments to accelerate the development of new and improved traits in corn, cotton, and oilseeds. . . . This decision was reached after a comprehensive review of Monsanto's research investment priorities and extensive consultation with customers in the wheat industry."[80] In short, the company had decided to write off its investment.

Understanding Monsanto's Decision

Monsanto's enormously costly and difficult decision to shelve GM wheat was no doubt linked to the pressure that the U.S. and Canadian anti-GM wheat campaigns had brought to bear on the company, coupled with widespread farmer reticence to risk the loss of overseas wheat markets. However, activist pressure and farmer resistance were not the only reasons. The tremendous beating that Monsanto had taken in Europe, the longer-than-expected time it took to bring many of its products to market, and the massive debt load the company acquired when it went on a seed company buying spree in the 1990s created a perfect storm of economic woes for the company.[81]

Reflecting these woes, Monsanto's stock price fell by half in 1999. That same year, the industry giant was compelled to merge with Pharmacia & Upjohn, a large pharmaceutical company whose motivation was to acquire Celebrex, a popular analgesic owned by one of Monsanto's drug units (Searle). Uninterested in the agricultural side of the business, which it viewed as a financial liability, Pharmacia spun Monsanto off into a separate agricultural company and sold 20 percent of it. Chief executive officer Robert Shapiro was shown the door, and a twenty-four-year Monsanto veteran by the name of Hendrik Verfaillie was given the job of trying to rescue the business.

Over the next several years, Monsanto's financial situation remained dire, and the company found itself in a make-or-break situation. In 2002, the company lost $1.7 billion, and by early 2003 its stocks had fallen to a mere eight dollars per share from over eighteen dollars a share in mid-2001. In 2002, Pharmacia sold Monsanto off completely, and the agricultural biotechnology business was left on its own, with no deep drug company pockets into which to dig.[82] Verfaillie was replaced by a new CEO, Hugh Grant, who sought to save the company by bringing in his own small-team-oriented management style and by accelerating the process of narrowing Monsanto's R&D priorities. It was in this context that Grant decided to halt the flow of money into Roundup Ready wheat and several other company research programs in 2004 and to concentrate Monsanto's R&D investments on those crops in which it had already achieved success (corn, soy, cotton, and canola). The company would no longer seek to develop GM crops that were "destined directly for the dinner plate," as one industry observer put it, but would "[focus] exclusively on seeds for agribusiness, ones that produced such goods as animal feed, ethanol, and corn syrup."[83] Ultimately, this decision paid off, and the company began to recover. By 2005, it was back on its feet, and by mid-2008 its stock price had climbed to over a hundred dollars per share.[84]

Movement Impacts

In letting Roundup Ready wheat go and focusing Monsanto's investments on genetically modified crops that had already been accepted by the market, Grant not only pulled the company back from the brink of disaster but also helped to deflate the opposition. Simultaneously with Monsanto's partial retreat from the dinner table came a decline in funding for anti-biotech activism. In the context of dwindling funds and the massive adoption of "first-generation" biotechnology crops by U.S. farmers,[85] the U.S. anti-biotech movement's energy fizzled. By 2004, the boom in activism had come to an end. While some U.S. activists remained dedicated to the cause, many had turned their attention to other issues. What remains of the U.S. anti-biotech movement in 2010, some three decades after its inception, is a handful of national organizations and a coterie of groups operating at the grassroots level. National-level organizations that are still active include the Center for Food Safety, whose executive director and core staff members have doggedly carried out legal challenges against agricultural biotechnology for thirty years; the Council for Responsible Genetics, which has also been in the struggle since the very beginning (and focuses mainly on human health, genetics, and biotechnology); and the Union of Concerned Scientists, which continues to focus on scientific, regulatory, and environmental issues around agricultural biotechnology. Grassroots groups have become active in a number of states, especially in Vermont and California, and have sought to develop local and state policies and ordinances that make it difficult to grow or sell GMOs in specific towns and counties. Occasional spurts of activism have continued to punctuate the industry's recuperation, but the movement has lost much of its momentum.

Even with this drop-off in activity, however, vestiges of the movement's impacts remain, both within the United States and abroad, where the U.S.-based activists have contributed resources and ideas to other anti-biotech movements. GM wheat and potatoes are not, as of this writing, available in the United States, and U.S. regulatory policy toward GMOs has become slightly more stringent. For example, certain safety test data that were once optional for biotechnology firms are now required by U.S. government agencies, although they still do not have to be made available to the public. The USDA also has a policy that requires farmers to plant non-GM crop "refuges" around their GM crops, in an effort to keep transgenes from spreading in unwanted directions. The refuge policy is a direct effect of activist pressure.

An *indirect* effect of anti-GM activism, we would argue, is also visible in the massive growth of the organic food sector and the growing concern about buying and consuming "pure food," at least among a segment of (generally more affluent) Americans. In the United States today, the only food that is at least nominally GM-free is certified organic food, the standards for which expressly prohibit the use of genetically modified organisms. Many American consumers who want to avoid GMOs in their diet therefore buy organic and shop at local farmers' markets, where farmers can personally tell them about the seeds they use and their growing practices. While teasing out the precise effect of anti-biotech activism on the demand for organics is not possible, there is a strong likelihood that it has been a contributing factor to this trend.

This activism may also be a contributing factor in the recent turnaround on rBGH. Beginning in 2007, a large number of U.S. milk processors started advising dairies that they would no longer accept milk from cows treated with rBGH. The apparent reason is that supermarkets are responding to consumer demand, which over the past decade has moved heavily in the direction of rBGH-free and organic milk. While the anti-rBGH activism of the 1980s described in this chapter is unlikely to have played a direct role in triggering this shift, it did help in creating a *labeled* "rBGH-free" alternative and in educating consumers about it. During the Clinton era, as consumers became both more health conscious and better off, they started to take advantage of this alternative, sending a strong market signal to milk suppliers. Ironically, some fourteen years after Monsanto introduced its rBGH product, Prosilac, demand for it is effectively dead. In August 2008, Monsanto sent ripples through the dairy industry when it announced it was exiting the milk hormone business and sold its rights to Prosilac to Eli Lilly.

CHAPTER SIX

Biotech Battles and Agricultural Development in Africa

To many observers, the pivotal moment in the global conflict over agricultural biotechnology was the closing of European markets to GM products in the late 1990s, engineered by anti-GM activists. These observers argue that it was the potential loss of lucrative European markets that made governments, producers, and traders around the world sit up and take notice. For instance, Robert Paarlberg has argued that African governments were not concerned about transgenic technologies until opposition exploded in Europe. Only then did they succumb to a rather inchoate (and unsubstantiated) fear of the technology's potential impact on health, environment, and trade relations.[1] Indeed, for many observers, the global reverberation of this activist "success" in Europe was marked most profoundly by the decision of several central African countries in 2002 to reject emergency shipments of World Food Program food aid on the grounds that it contained GM corn from the United States. Not surprisingly, this decision set off a firestorm of overheated rhetoric, as well as diplomatic wrangling, on both sides of the GM debate. Proponents, including the U.S. Agency for International Development (USAID), argued vehemently that the African governments were recklessly and baselessly endangering the lives of their poorest citizens by rejecting food that Americans had been safely consuming for over five years.[2] Critics of the technology countered that USAID and the World Food Program (WFP) were recklessly and immorally using hunger to launch the seeds of their imperial project on the African continent by refusing to tap non-GM sources of food aid. Many analysts noted these African governments' very real concern that they risked losing valuable export markets for their agricultural commodities if those commodities became "contaminated" with GMOs, given the powerful opposition to the technology in Europe. In late 2008, the editors of the prestigious British science journal *Nature* lamented "what have become known as Africa's GM wars," fomented by "European environmental groups and not a few African political leaders, for whom multinational businesses evoke the spectre of colonialism."[3] Thus, it appeared that the heated battle over biotechnology raging within Europe and between that continent and the United States had diffused into the poor countries of the global South, with profound ramifications for their citizens.

To be sure, the arguments and recriminations that swirled around the 2002 central African food aid crisis played an important role in politicizing transgenic technology on the African continent even before GM seeds and products could be legally imported, sold, or produced anywhere on the continent outside South Africa. Thus, the food aid crisis indicated clearly that the technology and the struggle over it had become global phenomena. Yet, as we show in this chapter, to focus attention on the cultural politics of consumption in Europe, or on the "successes" of anti-GM activists there, or on the opposition of African political leaders cannot fully capture the dynamics of contention over biotechnology in Africa, or indeed the processes by which specific political meanings came to be attached to GMOs by local proponents and opponents of the technology. In fact, African governments took a range of positions on the adoption of GM technologies, but beyond Zambia and Mozambique none unequivocally opposed it. Indeed, many African governments expressed the fear that if they rejected GMOs they would once again miss out on the benefits of a powerful new technology, just as they had with the first Green Revolution. Moreover, even in countries where governments were strongly in favor of transgenic technology, such as South Africa and Kenya, the issue became mired in ongoing controversy and contention. This conflict dramatically slowed the trajectory of the much heralded GM revolution and ultimately led agricultural policymakers and planners to reassess whether GM technology really was the key to solving African problems of productivity and poverty.

From early on, South Africa was an aggressive proponent of GM technology. In 1997, only a year after the first commercial plantings in the United States, it approved release of Monsanto's Bollgard *Bt* cotton, the first GM crop to be commercially planted in Africa. This was followed by *Bt* yellow and white maize (in 1998 and 2000, respectively) and by herbicide-tolerant cotton and soy in 2001.[4] Yet this enthusiastic uptake was not repeated elsewhere in Africa. Not until late 2008, when Burkina

Faso approved *Bt* cotton, did any other African country legalize the production of GM crops. Indeed, even then only a few African countries had constructed viable regulatory frameworks for biosafety. A new multilateral initiative aimed at producing a "new Green Revolution" in Africa, funded mainly by the Gates Foundation, had taken a cautious approach to GMOs, placing its faith mainly in conventional plant breeding techniques while keeping genetic engineering as one option. In short, in the face of sustained controversy, conflict, and challenge, the GM revolution had become bogged down in the very part of the world where its promised benefits were most direly needed, and the global trajectory of the new agricultural biotechnologies had shifted substantially.

As we show in this chapter, activists working on both local and global scales played a significant role in producing this outcome in Africa. African anti-GM activists participated in a range of transnational social and professional networks and drew on transnational discourses of development and imperialism, modernity, risk, and rights. But this did not mean that the impact of these social actors simply involved a diffusion of influence or resources from activists in the global North to their affiliates in poor countries. The social and political *meanings* that came to be attached to the technology were inevitably local, because local actors—regulatory policymakers, plant breeders, seed companies, extension agents, farmers, critics, and activists—interpreted and evaluated these technologies in terms of the specific local ecologies and local markets in which they were to be deployed. In helping to shape these local meanings, activists turned transgenics into an essentially contested technology.

Social activism affected the trajectory of agricultural genetic engineering in three interrelated ways. The first was through the role activists played in shaping the global biosafety regime, most notably in the multilateral negotiations over the Cartagena Protocol of the UN Convention on Biodiversity, an international agreement for governing trade and cross-border movements of GMOs. Working in the manner of transnational advocacy networks, activists from both the global North and the global South coalesced, lobbied negotiators and policymakers from potentially sympathetic states, and helped to craft the language of the protocol.[5] These activities played a crucial role in building a North–South coalition in the negotiations, which were able to push through a precautionary GMO trade regime. Moreover, the Cartagena Protocol provided activists and civil society watchdogs with a yardstick by which to measure governments' commitment to the international obligations they had accepted. They aggressively used this yardstick to press their governments to come into closer compliance with the requirements of the protocol whenever they showed an inclination to loosen regulation.

Local activists also shaped the trajectory of agricultural biotechnology in Africa by mobilizing evidence of "problems" with GMOs elsewhere, which they invoked to keep the shadow of doubt hanging over the technology. Most of this evidence was generated by activist–critics of the technology who were part of their global networks. Celebrated cases, such as the Starlink scandal in the United States, the Cornell study on monarch butterflies, the apparent "contamination" of traditional maize races in Oaxaca, Mexico, and the patenting of what Canadian activist Pat Mooney had dubbed "Terminator technology" provided African activists and civil society groups with powerful grist for their anti-GM and pro-precaution mill. These global reverberations were especially important because they made it more difficult for proponents of the technology to control the political field. For instance, though the Starlink scandal disappeared rapidly in the United States and Terminator technology was never commercially deployed, local anti-GM activists were able to keep these specters looming over GMO deliberations.

In their efforts to sow uncertainty around biotechnology, activists were helped by the massive defunding of agricultural R&D in most African countries since the 1980s, which had eviscerated the knowledge banks of these countries. Not only had qualified personnel become more scarce (partly because of the reduction of university budgets), but also a brain drain to the global North and to the private sector had occurred. There was also a higher turnover of scientists working on biotechnology.[6] In the context of a very real vacuum in authoritative *local* knowledge about biotechnology (including evidence of its benefits), the efforts of activists elsewhere became a powerful resource for local critics. By keeping this information in circulation and publicizing it widely, local anti-GM activists were able to maintain an alternative discourse about transgenic technology that ran more or less parallel to the proponents' discourse, with little intersection between the two and quite similar levels of popular authority.

The third and most significant impact of activism on the trajectory of GM technology in Africa rested on activists' ability to increase public scrutiny of governments' governance strategies over agriculture and technology. Adopting the role of public watchdog and calling for greater public oversight, activist organizations pushed African governments to proceed slowly and cautiously with the construction of mechanisms for biosafety regulation. A significant irony is at play here. Since the 1980s, international donors and financial institutions had squeezed African governments through structural adjustment programs and aid conditionalities to reduce their managerial role in the economy, to cut their budgets, to shrink in size, and to become more transparent. They had encouraged the emergence of an active and extensive NGO sector to take over tasks such as service delivery, social protection, microfinance, and development advice. As state funding of agriculture diminished across the continent, some of these civil society organizations took over agricultural development tasks, thereby establishing a presence on the ground that made them significant public interlocutors on GMOs and helped them to stymie the desires of the very neoliberal agencies that had encouraged their development.[7]

In sum, the dynamic interplay among activism at the level of global regime making, activism elsewhere (and for different local purposes), and local conflicts over the regulation of biotechnology enabled Africa's biotechnology opponents to keep the political pot of biotechnology boiling in a way that significantly slowed down its deployment. In doing so, they helped to push African governments and international agencies to reassess the technology's role in restoring agricultural productivity and resolving the African crisis of hunger and poverty. It is highly doubtful that anti-biotech activists would have enjoyed any success had one of these components been missing. It is also unlikely that they would have been able to make inroads had they confronted strong, coherent, and committed opponents who were able to convince the public of the social value of this technology.

In the remainder of this chapter, we first look at how anti-GM activists from the global North and South together worked to help shape the negotiation of an international biosafety protocol at the turn of the century. This international regulatory regime later became the basis for slowing down the deployment of GMOs in many African countries. We then turn to the ways in which local activists mobilized evidence generated abroad and circulated by activists from afar to help raise questions and concerns about the wisdom of adopting GM crops in Africa. In some countries, the evidence these activists mobilized dovetailed with their own governments' concerns about the threats that GMOs could pose to domestic agricultural exports and biological diversity. In other countries, governments took a strong pro-GM stance and were better able to deflect activists' concerns and to counter their narratives of risk with an alternative narrative of hope (and promise). Last, we focus on one strongly pro-GM African country (South Africa) to show how local activist groups, even in a context highly unfavorable to them, mobilized regulatory challenges that pushed the government to proceed with greater transparency.

Neoliberal Internationalism and the Biosafety Protocol

Many analysts have noted that in the era of modern globalization the international system has become more densely integrated in a variety of ways. A rapid increase in intergovernmental agreements and multilateral treaties has occurred as governments have sought to reduce transaction costs and resolve coordination challenges in a widening variety of issue areas, including public health, international migration, environmental management, and trade. In addition, supranational organizations such as the United Nations, the European Union, and the World Trade Organization have gained independent rule-making capacity. And the number, visibility, and influence of subnational and non-state-based organizations such as private corporations, nongovernmental organizations, and social movements (including terrorist networks) have expanded dramatically. Indeed, many analysts have argued that the role of states and the nature of sovereignty in the international system have shifted as power and agency have leaked both upward toward multilateral organizations and downward toward nonstate actors, such as multinational corporations.[8] The particular ways these developments have taken place in a specific issue area structure the transnational opportunity space for activism.[9]

With respect to agricultural trade and technology policies in which debates over the meanings and value of transgenic agricultural technologies are embedded, the transnational opportunity space for anti-GM activism was shaped in the 1980s and 1990s by the North's Globalization Project, which aimed to secure a regime of global economic governance that focused on the World Trade Organization. As we saw in chapter 1, the key aspects of this project involved liberalizing the agricultural markets of developing countries by reducing import quotas, tariffs, and other barriers to trade, "harmonizing" national regulatory environments by locating regulatory and rule-making authority in multilateral organizations, and strengthening private property rights through the WTO's Agreement on Trade-Related Aspects of Intellectual Property Rights (TRIPS). This project reflected dramatic power disjunctures between North and South, and African countries tended to occupy a particularly weak position because of their generally low levels of managerial capacity and their generally high levels of poverty and indebtedness.

As we noted in chapter 1, the globalization project was also accompanied by the rise of transnational activist networks organized around issues of global social justice and sustainable development. The activists within these networks helped to shape the political opportunity space of the biotechnology struggle by engaging both the interests of weak states and the international institutions involved in the globalization project. One of the most significant issues that they engaged was the international negotiation process to construct a global regulatory regime for cross-border flows of transgenic organisms under the Convention on Biological Diversity. The CBD was a product of the 1992 United Nations Conference on Environment and Development (the "Rio Summit"), at which environmentalists from both the North and the South had coalesced around the concept of "sustainable development" to develop common critiques of the Globalizaton Project's technology-driven agroindustrial development paradigm. Indeed, the Rio Summit had provided an extremely important catalyst of coalition building and networking among the rapidly increasing number of environmental groups and organizations that had emerged around the globe during the 1980s.[10]

During negotiations for the CBD in 1992, participants from the global South pointed out that their rich stores of biological diversity were uniquely vulnerable to the introduction of genetically engineered organisms. For this reason, they saw an urgent need for an international protocol to oversee global trade in GMOs, which they expected to increase dramatically in the coming years. But the United States and a handful of its allies perceived such a protocol to be a threat to the right of northern farmers and corporations to produce and sell their goods, including genetically modified agricultural commodities, worldwide and without prejudice. For the next eight years, they actively sought to prevent such an agreement from being negotiated.

Ultimately, the United States and its allies failed in their efforts, and the Cartagena Protocol on Biosafety was officially adopted in January 2000 as a supplementary agreement to the CBD. Although the final agreement was not as strong or comprehensive as most of the South wanted, it was broadly perceived as a victory by southern negotiators and the anti-biotech movement, both of whom sought *some* form of control over GMO trade. The protocol called for governments to adopt a precautionary approach in designing mechanisms to regulate biotechnology and to carry out stringent risk assessments of it. Moreover, although it required regulatory decisions to be made on the basis of "sound science," it allowed parties to use socioeconomic considerations in reaching decisions on importing GMOs.[11] Thus, it provided governments with a wider range of instruments to restrict trade in GMOs, without violating trade rules, than GMO proponents desired.

The Role of Activists

The successful negotiation of the Cartagena Protocol rested on a coalition of interests among anti-biotechnology activists, environmental activists, and governments from the global South. Anti-biotechnology activists sought a strong commitment to the precautionary principle. Environmental activists stressed the threats posed by new technologies to biodiversity and farmers' livelihoods in fragile agrarian systems. And poor governments were eager to assert their sovereignty against the neoliberal impulses of the international development regime. Although many factors contributed to the success of the negotiations, the role of social activists was crucial.

Prior to the negotiations, European activists and publics had put a tremendous amount of pressure on their governments to take the issue of genetic engineering seriously and to adhere to the precautionary principle. This pressure induced the EU to dispatch a number of high-level ministers to the negotiations, sending a clear signal to other parties that Europe intended to be responsive to its citizens. Representatives from a wide range of northern and southern NGOs also attended the meetings and used a variety of tactics to pressure delegates to reach a meaningful agreement. They engaged in a division of labor, sending some members into the meetings as formal observers and spokespersons and others out to the streets to demonstrate. Activist groups organized all-night vigils in the freezing Montreal weather, trotting out one colorful image after another to keep the media's attention. Apparently, they were quite successful. As one reporter for a Montreal newspaper noted, the activists "played the media like a Stradivarius."[12]

Activists on the inside also worked around the clock, lobbying delegates and disseminating up-to-the-minute economic, legal, and scientific information. Shortly after the negotiations began, the activist network released a position statement signed by twenty-five influential NGOs from around the world, demanding a strong biosafety protocol and admonishing those countries who were "only interested in protecting their industries and commercial investments in [genetically engineered organisms]" for obstructing the negotiations. Activist organizations also sought to shore up the power of southern delegations. Once they clearly understood that the United States had no interest in listening to the activists' position, they focused their efforts on forming a transnational alliance with state actors from the main negotiating bloc for the South, known as the Like-Minded Group. They supplied these southern delegates with information, helped them make rapid assessments of proposed text, and made concrete recommendations about specific language. They also brought environmental scientists to the meetings to give public presentations and invited the delegates to attend.[13]

Reflecting on the negotiations, Tewolde Egziabher, head of Ethiopia's environmental protection agency and the chief negotiator from the South, highlighted the importance of the activists' support when he explained how the Like-Minded Group, composed largely of representatives from African countries, managed to achieve anything against foes as powerful as the United States and its allies. "We had friends," Egziabher wrote.

> Africa is financially so poor that the African Group would not have functioned . . . without friends. But we soon made friends who filled in our gaps. . . . The Third World Network gave us critically needed assistance, and facilitated critically needed interactions, both South–South and South–North. African telecommunications are so poor that had it not been for the Gaia Foundation of London acting as an information relay station, we could not have been effective. And had our many, many other friends all over the world not helped, we would not have managed to stay as informed and as effective as we did.[14]

As one observer put it, "When you stand back, the key role the activists played is they were the eyes watching for society at large, calling delegates on the carpet if they didn't remember that they have citizens back home who might not be happy if they do certain things. . . . They were the watchdogs."[15]

Indeed, as we show below, activists have continued in that role, by using the Cartagena Protocol to demand that their governments live up to their international obligations. They have also appealed to the protocol as a way of pressing national regulatory agencies to widen the scope of public participation and social criteria in their risk and impact assessments for GMOs.[16] Consequently, as Gupta and Falkner have noted, "the process of [the protocol's] negotiation and implementation has created greater awareness of biosafety concerns and has strengthened domestic constituencies pushing for greater caution in testing and commercialization of biotech products."[17] In Africa, where thirty-seven countries have signed the protocol, government officials have found it repeatedly necessary to explain publicly how their attempts to construct regulatory regimes are consistent with its tenets. Thus the Cartagena Protocol has become a key reference point in national political conflicts over transgenic technology.

Contention in the Transnational Opportunity Space:
The 2002 Food Aid Crisis

The response of African governments to genetically engineered food aid should be interpreted in the context of this transnational opportunity space. African governments supported the CBD in part because it provided domestic regulatory instruments on risk management and intellectual property rights that held out some hope of holding the deregulatory pressures of the WTO at bay.[18] Consequently, these governments were inclined to stress the need for effective regulation, and they were inclined to invoke the precautionary principle. Moreover, in 2002 African governments had reason to be cautious about GMOs, because current evidence of their environmental impacts was uncertain, mixed, and highly politicized. For biotech critics and concerned governments, evidence of the technology's uncontrollability seemed ample, much of the evidence kept in circulation by activist organizations. One key example that local critics invoked was the 2000 Starlink controversy that activist organizations had engineered in the United States. Even as the World Food Program insisted that GM foods were absolutely safe to consume since U.S. citizens had been eating them for five years, critics noted that this was not true across the board, since some varieties (such as Starlink) were *not* approved for human consumption. When Starlink unexpectedly turned up in American food supplies, these critics pointed out that if unapproved GM varieties could find their way into the U.S. food supply, how much more likely was it that African food aid would be similarly tainted?[19] This threat seemed especially pronounced given the WFP's insistence that it was impossible to separate GM from non-GM corn in its food aid supplies (an insistence designed to convince African governments that food aid was an all-or-nothing option). Thus, even though the Food and Agriculture Organization and the World Health Organization both supported the WFP's claim that GMOs were safe to consume, this was not an argument that could be sustained *tout court*. Under these circumstances, the leaders of Zambia, Zimbabwe, Mozambique, and Malawi all asserted that there was no proof or even scientific agreement that these commodity crops were safe for human consumption; they were inclined to play it safe.

The potential threats to human health that GM food might pose were not, however, the overriding concern of these African governments. Indeed, with the notable exception of Zambia, they all ultimately agreed to accept GM food aid provided it was milled before crossing the border. Of much deeper concern to them was that some of the grain imported as food would inevitably end up being planted and thus "contaminate" local varieties and farming systems. Again, local critics drew on foreign examples, noting that in 2001 government scientists in Mexico had discovered GM corn being grown in fifteen areas of Oaxaca, even though the production of GM corn had not been approved in Mexico. (All GM corn in Mexico was imported for consumption or processing under NAFTA.) No one knew either what the vectors of contamination were or what the long-term environmental effects might be. But the case seemed to indicate that GM corn imported for consumption would inexorably end up being planted locally or otherwise make its way into the farming system.[20]

For poor African governments, as well as for African farmers, this broad concern contained two specific (albeit unmeasurable) threats. One was a perceived danger that, once introduced, GM varieties might squeeze out local varieties, reducing biodiversity and increasing the vulnerability of farmers to new, more resistant pests and weeds. This threat was regarded as especially serious for African farming systems, which were dominated by small-scale farmers working in conditions of high ecological variability. The second danger that agricultural practitioners perceived was that farmers would end up with patented seeds that they would be prohibited from replanting under the intellectual property rights regime promoted by the WTO.[21] In most of Africa, plant breeding is still a fluid process, with farmers steeped in traditions of seed saving, farmer-to-farmer exchange, and locally specific varietal selection as a key tool for responding to expected but unpredictable ecological or climatic shocks. For these farmers, the vast majority of whom are small scale and resource poor, this concern ran deep. Moreover, it was intensified by fears that, as multinational companies established control over local seed supplies, they would impose the so-called Terminator technology on farmers, thereby tightening the grip of dependency.[22]

To many in the development community and aid agencies, these fears were inchoate, unfounded, and irresponsible.[23] Some saw the African governments' decision as driven entirely by a fear of losing access to European markets and therefore determined ultimately by the self-indulgence of well-off European consumers who had no need to think about the impact of their consumption decisions on the distant poor. Others thought it showed the undue influence of mischievous and unaccountable environmental groups, mainly from the global North, whose main interest was to augment their power. Yet to appreciate the politics of GM food aid, we should also recognize the potent institutional vulnerabilities of these countries. Zambia and Malawi were among the poorest countries in Africa, with respective GDPs per capita of $780 and $570. Both were locked in stringent debt reduction programs with the IMF that required them to constrain government spending tightly. They also relied heavily on the export of primary goods for revenues. In short, these countries had very limited options for securing the livelihoods and well-being of their citizens over time.

Arguably, for these countries, the central concern catalyzed by the food aid crisis was not the indeterminate threats to health or environment but rather their own poor capacity to govern the agricultural sector. As Chinsembu and Kambikambi put it, in the case of Zambia,

> the extension services and education system lack the capacity and trained personnel to bring farmers up to date with developments in agricultural biotechnology, and genetic engineering in particular. There are no serious awareness campaigns to inform stakeholders about this new technology and even at university level there are no courses in biotechnology. The media has stepped in to fill the gap, but many journalists have no access to reliable information. The internet is unreliable or non-existent even for those working on the major national newspapers and journalists lack contact with specialists who could keep them in touch with what is going on in the region and internationally. This has created a situation where the agricultural sector as a whole is vulnerable to misinformation and the opinions circulated by those with [hidden] agendas.[24]

Other than Zimbabwe, none of these countries had adequate biosafety policies or effective frameworks for intellectual property protection in place. Thus, the food aid crisis confronted them suddenly and directly with a new sense of vulnerability: not only did they lack the institutional capacity to control the movement of GM seed across and within their borders, but they also had no effective legal frameworks for adjudicating the rights and responsibilities of plant breeders, seed companies, farmers, and consumers. The impact of this realization was pointedly captured in a "living paper," "Governing Biotechnology in Africa," presented at the 2004 meeting African Policy Dialogues on Biotechnology, held in Zimbabwe:

> The presence of GM food in the region did not only raise political temperatures, it also rendered inordinately difficult a range of other basic tasks and operations in food relief—such as moving grain through ports and across borders. Perceived risks associated with GM food created an entirely new set of transactions costs. How, for instance, in mid-2002 was Malawi to move maize donated by USA,

and thus containing *Bt*-maize, through Tanzania in the absence of complementary Biosafety Protocols in Tanzania and Malawi, and in the absence of associated testing machinery? Ad hoc measures had to be hammered out . . . on such seemingly mundane issues as: how to load grain into rail cars and trucks with minimal "escape"; how to cover the loaded trucks and cars; how long to allow the loaded cars and trucks to sit in given positions. The opportunity cost associated with such logistical hurdles, coupled with the region's general reticence towards potential life-saving but GM food, elicited intense scrutiny and opprobrium from food donors and relief agencies.[25]

In effect, the food aid crisis brought these governments face to face with the power and implications of the WTO agricultural governance regime, not least because the GM corn was provided by the United States—a nonsignatory of the Cartagena Protocol—with absolutely no regard for the regulatory strictures that applied to participants.[26]

Under these circumstances, the risks of crop and farmland "contamination" seemed extraordinarily high. In the first place, African governments feared the loss of potential international markets, not only because this would result in lost revenues, but also because, in an era of globalization, a country's ability to certify the standards, properties, and qualities of its traded goods is an important marker of its trustworthiness as a trading partner. The failure to demonstrate such capacity could threaten a country's ability to enter into new or future international markets, let alone existing ones.[27] Poor African governments had good reason to fear that their lack of a clear and effective governance regime for the technology might incite prejudice against them in international trading relations. As a spokesman for the Zambian Department of Agriculture explained his country's decision, "The Zambian government does not have the capacity to detect whether food is genetically modified; we have not yet ratified the Cartagena agreement and we have no legislation in place on biotechnology and biosafety."[28] Ironically enough, African governments' sense of insecurity was heightened by the knowledge that international activist networks would eagerly expose such contamination.

Consequently, the crisis constituted a wake-up call for African governments to focus more attention on devising their regulatory frameworks for biotechnology, including the protection of biodiversity resources.[29] It also put pressure on African governments to harmonize or coordinate their regulatory frameworks, because their agricultural industries regularly moved seed and grain across intra-African borders. As the ongoing wrangling within the European Union over GM regulations demonstrates,

such harmonization is always difficult to achieve in international relations, even when those relations are highly institutionalized.

In the second place, African governments feared the contamination
of their land by what they considered an uncertain technology and its
potential impact on agricultural productivity. Given the established tradition of improving crops through farmers' on-farm experimentation in
collaboration with public research institutions, even the smallest threat
of losing that capacity to improve was frightening to a poor government
whose plant-breeding R&D capacity was tied up in such techniques. This
was particularly true because public funding and institutional capacity
for agricultural R&D had been declining steadily for two decades. This
decline was partly the result of the austerity measures imposed on heavily indebted countries by the international financial institutions, which
had pushed governments to reduce funding of agricultural research and
development.[30] Another reason for this decline was the marked trend
toward liberalizing the agricultural sector during the 1990s, which led
many African governments, under pressure from international donors,
to privatize their parastatal seed supply and marketing operations. These
operations had been a source of revenue for agricultural R&D funding.
Meanwhile, as the governance of agriculture slipped further from these
governments' grasps, under-resourced scientists needed to be able to do
their work, and local crop strains had to be protected.

In short, and quite ironically, central and southern African governments saw in GM crops a profound and multifaceted threat to their sovereign capacity to feed their populations. More particularly, risk assessment
frameworks based on "sound science" had to be constructed under conditions of complex uncertainty. Many proponents of transgenic technology
saw this uncertainty not only as an opportunity but also as an urgent
imperative to move this scientifically cutting-edge technology ahead rapidly for the public good. Critics saw it quite differently: as a profoundly
disempowering technology that exposed poor countries and marginalized
people to potentially devastating uncertainties and risks and to the imperial domination of patents. These risks, they believed, mandated a total
rejection of GM technology for the public good.

Science, Uncertainty, and the Politics of Regulation

The politics of food aid in central Africa was not driven directly by activists, though activists weighed in vociferously at both the national and
the international level. In fact, the way the politics of food aid played out

shows how activists' efforts helped to fashion a transnational opportunity space for negotiating agricultural biotechnology in which the technology became inherently politicized and controversial; the very use of the term *GMO* invoked fears of health, environmental, and economic risk that influenced the behavior of governments, farmers, and NGOs, as well as some consumers. Partly because African countries were largely "technology takers" rather than "technology makers" (i.e., they depended more on the diffusion of technology than on the production of it, including seed technology), their principal concern focused on how biotechnology could be effectively governed and how they could assert their national sovereignty over those who controlled it. Consequently, the forces and dynamics at play in the transnational opportunity space for activism focused the politics of anti-biotech activism on the *regulation* of biotechnology and of the mostly foreign interests that drove its deployment in Africa. And, perhaps inevitably, the battles over biotechnology were most heated in those countries where agricultural technological capacity was fairly well developed, and reasonably robust regulatory mechanisms for biosafety were already in place by the time of the 2002 food crisis, countries such as South Africa and Kenya. Unlike most African countries, these countries did have some scientific capacity to get in on the GM revolution and could hope to be serious (if perhaps small) players in the global biotechnology industry.

It is noteworthy that even though transnational activism had opened up the regulatory system as a space of contention, the regulatory realm is not a space that is particularly congenial to absolutist claims: it is a space in which the conditions of deploying the technology could be negotiated rather than any question of its adoption. Thus, GM proponents could not claim categorically that GMOs are safe, given that some were not approved or were approved for only limited purposes. Similarly, critics could not claim categorically that GMOs are unacceptable given that some met regulatory requirements. For each side, therefore, the trick was to move the regulatory framework as close to that side's absolutist claim as possible. In this context, technology proponents pressured African governments to adopt permissive regulatory frameworks, whereas activists and other critics pressured these governments to be more precautionary.[31] In particular, activists focused a considerable amount of energy on blocking or tightening the regulatory process in order to keep GMOs out of the public realm.

In each of these cases, conflicts over biotechnology were cast in terms of two clashing narratives about the value and meaning of this technology.

On the one hand, proponents invoked a narrative of *hope,* in terms of which transgenic technology exuded the promise of raising agricultural productivity levels to reduce poverty and to meet the food needs of a rapidly increasing world population.[32] On the other hand, opponents invoked a narrative of *risk* and *rights,* laying particular stress on the uncertainty of the technology's long-term impacts on the environment and human health, the threats to farmers' control over their farming system, and the right of consumers to choose what they eat. In doing so, activists adopted the frames of their allies in the global North and recast them according to local histories, cultures, and institutional opportunity structures. In the remainder of this chapter, we look more closely at the dynamics of these contentions in South Africa.

The South African Story

As noted above, South Africa has been the African leader in pursuing transgenic technologies as a scientific and agricultural development strategy. It was the first African country to formulate a regulatory framework for biosafety based on the Genetically Modified Organisms Act, which went into effect in 1999. It had a long tradition of public agricultural research, several lively research institutes that were already internationally networked, and considerably more research capacity than other African countries (except, perhaps, Nigeria, Egypt, and Kenya, with whose research establishments South African scientists were developing increasingly strong links). As Ofir noted in the mid-1990s, "It is expected that South Africa will play a major role in the development of human resources in biotechnology elsewhere in Africa, directed towards the empowerment of African scientists on the continent to serve the needs of their own countries."[33] In addition, South Africa had a stronger regulatory capacity than most other African countries. It was one of very few to have statutory plant breeders' rights and was a member of the International Convention for the Protection of New Varieties of Plants (UPOV).[34] Consequently, South African authorities and other regional politicians expected that South Africa's regulatory framework for biotechnology would provide the model for other African countries still formulating their own policies.

The Regulatory System

The history of transgenic agricultural biotechnology in South Africa goes back to 1989, when its Department of Agriculture received its first application for a GM field trial of *Bt* cotton. Lacking a biosafety framework at the time, the government authorized the trial under the 1983 Agricultural Pest Act and ceded the mandate for conducting scientific risk assessment and establishing field containment requirements to the South African Committee on Genetic Experimentation (SAGENE). Established in the late 1970s as the national advisory body on biotechnology research and development,[35] SAGENE was led by life scientists who were based in universities and research organizations and were pursuing their own individual research interests and projects but shared an enthusiasm for genetic and molecular research. Government science policy offered little support, because the apartheid government's research interests focused on specific outcomes in applied research, especially first-generation plant biology. Little coordination existed among research establishments or among government, academia, and industry. Thus, early biotechnology development was driven largely by the personal initiative and ambition of these individuals, who recognized that the gene revolution was rapidly changing the international field of genetic and molecular research and were concerned about the danger of being left behind professionally. These scientists established SAGENE mainly to develop sound science research protocols around such "bench science" issues as experimentation, safety, et cetera.[36] In doing so, they gave institutional form to a specialized "epistemic community" that was cogent but also exclusive and prescriptive.

With the advent of "new" biotechnologies in the early 1990s, as well as the shift in policy frameworks for science and technology associated with the end of apartheid, the research groups associated with SAGENE moved almost inexorably to the center of the regulatory process. They also built extensive international research networks, both with academic colleagues and with funding organizations, especially in the United States. As they attracted international research funding and institutional collaborations, they also strengthened local networks of science research and raised the profile of the local scientific community. Consequently, they built themselves a strong position to speak with authority on issues of science policy and, indeed, to place their own normative framework of authoritative knowledge, which stressed "good science" (laboratory-based scientific method), at the heart of policy debates. Institutionally, they expressed this authority through SAGENE, which played an advisory role to all government agencies on matters pertaining to the importation of GMOs or their release into the environment, or both. Indeed, SAGENE provided the guidelines for field tests of transgenic crops and was responsible for evaluating risk (regarding human food, animal feed, and environmental impact) of all applications requesting authorization to conduct GMO activities.[37] As the anti-biotech group GMWatch noted, "As the new South African government, which was ushered in on the 27th April 1994, had no particular knowledge or expertise in these areas, regulatory matters were left very much in the hands of SAGENE."[38]

The year 1997 turned out to be a big one for agricultural biotechnology in South Africa. The first commercial GM crops were planted, the Genetically Modified Organisms Act was passed "to provide for measures to promote the responsible development, production, use and application of genetically modified organisms," and the first anti-GM organization, Biowatch, appeared.[39] Promulgated with the task of *promoting* transgenic technology, the GMO Act was in fact a very permissive regulatory instrument. Since it was written well before the Cartagena Protocol came into effect, the act did not take the precautionary principle into account.[40] Rather, it was modeled on the U.S. regulatory approach, which stressed the concept of "substantial equivalence" and thus made no effective distinction between "old" and "new" biotechnology techniques in devising risk assessments. Moreover, as in the United States, the Advisory Committee, constituted under the GMO Act to replace SANGENE as the assessment agency, conducted safety reviews on risk assessments submitted by applicants rather than conducting those assessments themselves. The act applied only to living GMOs and not to products derived from GMOs, and it did not include any regulations for labeling GM products.

The permissiveness of the GMO Act was designed to facilitate both rigorous regulation and rapid approval of GMO applications. In order to fast-track approvals, regulators frequently regarded trials conducted in the United States as adequate for regulatory purposes, and evaluations carried out under the GMO Act were often desktop- rather than field-based.[41] This was regarded as important for efficiency and swift regulatory clearance. The executive director of AfricaBio (the most active pro-GM lobbying organization in the region) argued in a parliamentary public hearing in 2006 that South Africa should streamline its regulatory process: "There is quite an extensive amount of data that has already been collected globally," which, she urged, should be examined in a more comprehensive manner so that the country does "not waste [its] resources trying to re-invent the wheel by doing it all over again."[42] While the provision of safety checks was essential, she noted, also essential were their ease of use by both scientists and farmers and the minimization of costs to ensure maximum benefits from the technology.

If the regulatory system was permissive, it was also exclusive. The GMO Act placed regulation of GMOs within the very technology-friendly national Department of Agriculture, and its Advisory Committee, consisting of ten scientists who were experts in the field of GMOs, continued to draw heavily from the group that had been networked around SAGENE. Indeed, several of them played a key role not only in writing the National Biotechnology Strategy but also in drafting the GMO Act. Not surprisingly, this regulatory framework reflected the norms, methods, and standards of the epistemic community from which it came. Regulatory officials tended to see biotechnology regulation as a straightforward matter of science and technology policy and consequently drew the circle of expertise tightly, privileging research scientists and plant technologists, whether from the academy, the government, international consultants and foundations, or the commercial companies themselves. Social scientists, who tended to have a stronger presence in development circles, were scarcely included, thereby entrenching a perhaps invidious separation between the realm of agricultural biotechnology policy and the realm of development policy, in which social scientists played a significant role.[43] Moreover, although the GMO Act formally recognized farmers and consumers as crucial stakeholders, both groups were effectively frozen out. The Advisory Committee made no allowance for public participation, which was limited to a "notice and comment" procedure linked to permit applications for environmental release.

Most particularly, this regulatory system gave no consideration to the capacity of the state to provide (or even to design) appropriate resources, infrastructure, extension, and education to the farmers who would end up putting GMOs into the environment. This outcome was somewhat ironic, since the first regional seminar on biotechnology policy, organized by the Intermediary Biotechnology Service in 1995, had stressed the importance of orienting the technology toward small-scale farmers and integrating social sciences in policy planning. In fact, seminar participants—from foundations, national science and technology centers, industry, et cetera, but excluding farmers—had generally agreed on the importance of including farmers in the construction of biotechnology frameworks.[44] But when it came to the practice of regulation, small-scale farmers' voices were excluded. As one GM field trial application put it, "Emerging and subsistence farmers will be invited to the trials to view the technology, [to] ask questions and to be consulted on its potential value to them."[45] This exclusion was particularly important, since the act enshrined the principle of "end-user" liability for any harm caused by GMOs, rather than the "polluter pays" principle that was written into other major legislation, such as the National Environmental Management Act.[46] Thus, there was a built-in tension in this regulatory approach: even though the benefits of agricultural biotechnology could be realized only at the level of the farming system, the farming system itself could not constitute an *integral* part of its evaluation, and potential risks accrued substantially to the farmer.

The Emergence of Opposition

This promotional atmosphere for GMOs provided the context for the first GM approvals in South Africa in 1997, merely a year after they were first commercially produced in the United States and before the GMO Act had been finalized. At the same time, a small but vociferous network of activists began to organize against the development and deployment of GMOs in South Africa. Many of these activists were middle-class whites with long and overlapping personal histories of working in universities or in the NGO sector on social and environmental justice issues. Indeed, many had become involved in environmental politics during the struggle against apartheid, and this influenced both their views and their skills as activists.

These early environmental activists were not a cohesive group. Some came to the issue through urban campaigns that focused on the systematic exposure of poor and marginalized communities to industrial pollution, toxic waste dumping, and inadequate sanitation facilities. Others were more concerned about the ability of marginalized rural communities to access and use land sustainably. A third set focused on the socio-environmental effects of South Africa's nuclear program, which was developing both military and power-generating capacities. Despite this diversity, however, these various groups of individuals interacted frequently in the rising tide of civil society activism that characterized the struggle to end apartheid in the 1980s. They attended the same meetings, briefings, and protests. They shared a sense of environmental justice that was community-oriented and linked to social justice, and they were deeply sensitive to the history of dispossession linked to the development of racial capitalism in South Africa. They were therefore concerned about development choices that might further disempower the poor, and they were inclined to see threats to biodiversity as linked to social exclusion.

The focus of these activists on the rights of marginalized communities and their sensitivity to histories of dispossession under colonialism were important for two reasons. First, they gave these activists what one might call a "decommodification bias" in their analytical and normative frameworks. But perhaps even more important, they provided these activists with a strong solidarity with activists in other parts of the ex-colonial world, such as India and Malaysia. Over time, this group of South African activists fashioned networks of communication and solidarity that reached across local, national, and international spaces and provided a launching pad for anti-biotech resistance to emerge in the late 1990s. For instance, in 1992 (immediately after the Rio Summit) the environmental justice organization Earthlife Africa hosted a conference on sustainable development that included delegates from across the North and South, such as Kristin Dawkins, from the Minneapolis-based Institute for Agriculture and Trade Policy; Martin Khor, from the Malaysia-based Third World Network; and Vandana Shiva, from the Research Foundation for Science, Technology, and Ecology. All of these highly visible international activists were part of the transnational group that fashioned the intellectual architecture of the anti-biotech movement in the 1980s (see chapter 3). At the conference on sustainable development, the South African Environmental Justice Networking Forum was initiated. More localized groups, such as the Group for Environmental Monitoring, also emerged, as did community-based organizations, such as the South Durban Community Environmental Alliance.

As Jacklyn Cock has noted, there was little organizational, strategic, or even analytical coherence to these networks. Rather, they constituted "an inchoate sum of multiple, diverse, uncoordinated struggles and organizations."[47] Yet they brought key resources to the mobilizations against biotechnology. One resource was networking skills and contacts in a civil society that was already highly mobilized to press for extensive rights and powers of democratic participation in the postapartheid society. Another was a more specific demand for accountability and transparency in government agencies responsible for managing social and environmental resources. A third was an established ability to incorporate their concerns in the agendas of some government agencies (such as environmental agencies) and to participate actively in certain government policy deliberations. Finally, through their personal histories of activism, they had confidence and conviction that they could make a difference.

In 1997, alarmed at what they perceived to be a wholesale, inchoate, and untransparent process of approval for GM permit applications, some of these activists founded the first anti-GM organization, Biowatch South Africa, led by a biologist formerly at the University of Cape Town. Biowatch's brief was "to publicise, monitor, and research issues of genetic engineering and [to] promote biological diversity and sustainable livelihoods," as well as "to prevent biological diversity from being privatized for corporate gain."[48] Three years later, the South African Freeze Alliance on Genetic Engineering (SAFeAGE) was established with the aim of pursuing a broader public strategy that focused on public awareness, food safety, and consumer choice. Modeling its demands on the European Union moratorium on GMO deployments and stressing the precautionary principle, it called for a comprehensive five-year freeze on all GMO activity to allow more time to assess the health, safety, and environmental implications of genetic engineering if it was to be used in food and farming. SAFeAGE also pressured the government to ratify the Cartagena Protocol. Constructing the alliance self-consciously as a network and not as a campaign organization, the SAFeAGE leadership worked mainly to orchestrate coalitions of civil society organizations around specific activities, such as media events, information sessions, supermarket campaigns, labeling campaigns, petition drives, et cetera.[49] The alliance managed to secure the support of important constituencies, such as the Catholic Bishops Conference and the labor federation Cosatu, and it enjoyed generally supportive coverage by the media. Working closely with other local, anti-biotech organizations, SAFeAGE made a great deal of information public in an effort to secure the precautionary principle as a public demand and a regulatory norm.

A third organization, the African Center for Biosafety (ACB), was created in 2003 to push for the implementation of stringent and comprehensive biosafety policies, to increase the role of civil society in protecting African biodiversity and food production systems, and specifically to oppose the commercialization of GM crops. The ACB was directed by Mariam Mayet, an environmental lawyer who had previously worked for Greenpeace International and for Biowatch. The ACB took on a watchdog role, gathering information and disseminating critical analyses on developments in the biotechnology industry, at home and abroad. All seven members of its board had an academic background, and the organization prided itself on the production of careful research.

Over the ensuing decade, these three organizations took the lead in trying to beat back the behemoth of biotechnology. Broadly, they were motivated by a belief in three related sets of citizenship rights that they saw as threatened by the headlong rush to embrace agricultural biotechnology: the right to a safe food system, the right to a healthy and sustainable environment, and the right of citizens to choose (including not only the right of people to know what they are eating but also the right of farmers to select the seed varieties that they wish to plant). In short, they sought a different, and broader, notion of sovereignty over seeds—one in which public voice would play a larger role in assessing the risks of the technology, as well as its value, at least in part on the basis of informed citizen choice.

They believed that these normative commitments should be considered inseparable from knowledge claims regarding technological innovation and should be incorporated into the regulatory system. They were also convinced that the development and deployment of GMOs in South Africa had been captured by a very small group of scientists with little accountability, transparency, or oversight and that no effective, authoritative, or responsible independent monitoring of these developments was being provided, by the government or by anyone else. Most crucially, they tended to see the science of transgenics as inextricably tied to external private interests in ways that actively suppressed alternative ways of doing things, such as saving and sharing seed, conducting farmer-based research, and farming on a cooperative, organic basis.[50] In the view of these organizations, GMOs were a Trojan horse for international corporate control of South African agriculture, with the likely outcome of increasing levels of impoverishment among South Africa's smallholder farmers. Given their sensitivity to South Africa's histories of colonialism and apartheid, they viewed this threat of dispossessing the poor as profound.

These concerns brought them into close networking alliances with international organizations such as Greenpeace, the Barcelona-based GRAIN, and the Malaysia-based Third World Network, all of which had campaigns against GMOs, and all of which shared resources with the South African organizations. They also drew funding from the London-based Gaia Foundation, the Amsterdam-based HIVOS, and the Germany-based GTZ. Locally, they networked with consumer rights groups such as the Safe Food Coalition and the Consumer Institute of South Africa, as well as environmental rights organizations such as Earthlife Africa and the Organic Agriculture Association of South Africa. None of these groups was large or well resourced; in many cases, they consisted of a Listserv and two or three coordinators who came together irregularly to work on local campaigns.[51] But they drew strength and resilience from close personal ties, shared histories, and interlocking memberships that provided an organizational base for them to conduct more concerted campaigns. For instance, in 2002, SAFeAGE drew on both its international contacts and local networks to organize a number of public events on genetic engineering at the World Summit on Sustainable Development, in Johannesburg, South Africa, where they clashed loudly with GM proponents. In short, these organizations were able to catalyze a thickening network of global and local actors concerned about sustainable development, biodiversity, farmers' rights, et cetera and bring them to bear on the perceived threats of transgenics.

Nationally, these organizations worked hard to generate a public awareness, and suspicion, of GMOs, especially after 1998, when seed companies began to sell *Bt* cotton to smallholder cotton farmers in the Makhathini Flats region of KwaZulu-Natal—the first smallholder farmers in Africa to receive genetically modified seed—and when the government approved a genetically modified white maize, the first staple foodstuff to be approved worldwide. Activists saw these developments as highly dangerous, both to consumers and to small-scale African farmers because of the "new and unpredictable risks for both human health and the environment" that these crops and foodstuffs posed.[52] From their perspective, the government was plunging in recklessly with an irresponsible and woefully inadequate regulatory apparatus. Consequently, the vast majority of these organizations' strategies to beat back biotechnology targeted the arenas of policy and regulation.

South African anti-GM activists realized that the GM train could not be stopped unless the regulatory system could be brought out from behind the narrow regulatory shroud of "sound science" as defined by a small group of scientists and policymakers and opened up to greater public scrutiny, accountability, and transparency. Consequently, they focused their strategies principally on widening the definition of authoritative knowledge for risk assessment on each of the dimensions discussed above. They also sought to widen the risk-management criteria that should be included in regulatory decision making. One target was a demand for the comprehensive labeling of GM products.[53] Another target was a demand to broaden the criteria for environmental impact assessments to include socioeconomic factors. For instance, some activists argued that if the adoption of GM seed would require a farmer to implement non-GM "refuges" that would be difficult or expensive to set up, those costs should figure in decisions about whether to approve a particular seed; in other words, technology should not drive the farming system.

In pushing to open up the regulatory system, activists had access to two key resources. One was the Cartagena Protocol, which mandated environmental impact assessments, public participation in the regulatory process, and the acceptability of including social and economic criteria in technology impact assessments. South Africa had been a member of the Like-Minded Group during the negotiation of the protocol, and since the country expected to take the lead in the regional production and trade in GMOs, it had good reason to take the Cartagena Protocol seriously. Indeed, eager to show its credentials in the international arena, South Africa had succeeded in being elected to the Intergovernmental Committee on the Cartagena Protocol, which was set up to bring the protocol into effect. Yet the GMO Act had been put in place before the protocol was signed into effect in 2000, and many of the act's features were not in full compliance with the protocol. Caught awkwardly between two divergent predilections—one to be a responsible citizen with respect to participation in international institutions, and the other to lead in technology development—the South African government stalled on signing the Cartagena Protocol until August 2003.

Meanwhile, anti-biotech activists joined a broader panoply of environmental and other groups in pushing the government to sign the protocol and, thus, were able to keep the issue of biotechnology high on the government's agenda. Once the government did sign, officials recognized the need to harmonize a variety of policy instruments on environmental management, patent protections, plant breeders' rights, and so forth, including the biosafety framework provided in the GMO Act. Anti-GM activists kept up pressure on the government to meet this need and insisted on making expert and public submissions, attending public input and feedback sessions, and carrying out public lobbying campaigns, especially on such controversial issues as mandatory labeling of GMOs.[54] Thus, the Cartagena Protocol provided activist groups with an important ratchet for pushing the government to reformulate the regulatory framework for agricultural biotechnology and for increasing activists' own participation in the process. In so participating, they kept the debates around the amendment of the GMO Act—in particular, the question of labeling and the public's right to know—in the public eye.

The other set of resources on which they could draw to pursue these aims was the broader panoply of national regulatory mechanisms that could be invoked to apply to GMOs, thereby constraining the promotional exuberance permitted by the GMO Act. Particularly relevant was the National Environmental Management Act (NEMA) of 1998, which made extensive provision for public participation and public information related to activities that might have an impact on the environment. This act had been passed following years of deliberation that included substantial public involvement in which environmental justice activists had participated actively. As a result of this history, the national Department of Environmental Affairs and Tourism was much more open to civil society input than was the Department of Agriculture, which controlled the GMO Act. Since the Department of Environmental Affairs had substantial interest in the impact of GMOs on the environment, activists used their access to ensure that the more inclusive and precautionary approach of NEMA was included in the reformulated GMO Act.

In these efforts, activists were aided by the broader demands for democratic voice and public accountability that accompanied the transition from apartheid. In particular, they were able to monitor the government's decisions closely by appealing to the Promotion of Access to Information Act of 2001, which was aimed at increasing the transparency of government activities, and the Promotion of Administrative Justice Act of 2001, which required government agencies to explain their actions and decisions at the request of civil society organizations. Much of the battle over agricultural biotechnology in South Africa comprised the efforts of activist organizations to use these tools to push regulators in more precautionary directions.

This strategy is revealed most directly in two legal actions that Biowatch brought against the National Department of Agriculture. The first was a suit brought in August 2002 requiring the registrar of the GMO Act to provide Biowatch with comprehensive data on risk assessments, in terms of a "right to information" secured under section 32(1) of the Constitution. The second was an appeal that Biowatch lodged in April 2004 against the registrar's decision to grant Syngenta a permit for the import, field trials, and general release of Bt 11 maize. Biowatch won the first of these actions, but in a rather bizarre twist the court required it to pay the legal costs of Monsanto, which had intervened in the case as an interested party (and in particular wanted to protect information it had released to the government for risk assessment purposes but regarded as proprietary). The costs to Biowatch were especially high, because Monsanto's intervention had required the activist organization to commission an expensive affidavit to address Monsanto's concerns.[55] Biowatch lost the second case, partly because the organization was unable to muster the independent scientific expertise to undermine Syngenta's documentation.

With these actions, Biowatch had several objectives. One was to learn as much as possible about the actual process of evaluation under the GMO Act in order to devise effective counterstrategies. The second was to pry open the process for public scrutiny. The third was to try to inject the National Environmental Management Act, which requires both a "risk-averse and cautious" approach and participatory environmental governance, into the politics of biotech regulation. The Syngenta appeal demonstrated the importance of this last aim for Biowatch's efforts to open out the regulatory process in two ways. First, the registrar was fast-tracking the permit process by allowing simultaneous field trials and general release. This process not only undermined the very point of field trials but also was incompatible with a risk-averse and cautious approach to environmental management. Second, because Syngenta short-circuited the field trials phase, its risk assessment could not include significant local data. Instead, it relied on studies carried out in the United States and on data for a similar, though not identical, product. This approach to risk assessment, Biowatch argued, ran counter to the participatory governance requirements of NEMA and took inadequate account of the potential impact on *local* biodiversity. As such, it raised questions about the portability of scientific "facts" versus the role of contextual knowledge in evaluating technology transfer. This strengthened the organization's calls for a science-based approach to regulation that was independent, local, and broadly public.

As anti-GM activists learned to navigate the regulatory process, their constant pressure had a significant effect. In 2003, they helped push the government to acknowledge shortcomings in the monitoring and oversight of GM technology and to agree to amend the GMO Act in order to harmonize its relationship with the Cartagena Protocol and the National Environmental Management Act. While this process dragged on, activists kept the pressure on through repeated challenges to GM applications brought before the government. For instance, in June 2004, a coalition of activist groups filed a multipronged objection to an application by the parastatal Agricultural Research Council (ARC) for field trials on *Bt* potatoes. They complained that ARC had not made information available in a timely fashion for public participation in the regulatory process and that the project had failed to consider all the threats that the *Bt* potatoes could pose to the environment. They also complained that ARC had not met the statutory requirements of the Environmental Conservation Act, NEMA, and the GMO Act.

Especially noteworthy was their invocation of provisions under NEMA: (a) that a risk averse and cautious approach should be applied, taking into account the limits of current knowledge about the consequences of decisions and actions, and (b) that "the social, economic and environmental impacts of activities, including disadvantages and benefits, must be considered, assessed and evaluated, and decisions must be appropriate in the light of such consideration and assessment."[56] The potato project, the challengers argued, could not meet either of these legal requirements, not only because the broader environmental risks were unknown, but also because it was *"not possible"* for these potatoes to benefit the target group of poor, small-scale farmers, since, inter alia, the seeds were subject to ten patents, they were more expensive and would drive farmers into debt, they were not engineered for yield increase, and people probably would not eat them anyway.[57] This was a novel argument that sought to use existing legislation in order to inject a broader set of criteria into regulatory decisions that would change the calculus of this technology's *inherent* value and meaning. Inasmuch as some of these criteria required an assessment of the probability that the technology would alleviate poverty, they were quite speculative. Nevertheless, this challenge demonstrates activists' efforts to push regulatory norms in more precautionary, inclusive, and transparent directions.

The *Bt* potato challenge was but one riff in a drumbeat of such challenges that fit an activist pattern one might sum up as "scrutinize, publicize, oppose." Biowatch and ACB, in particular, followed applications closely and raised a red flag over every one. In 2004, it was ACB that challenged an application by Monsanto SA for food and feed safety clearance for Roundup Ready wheat, even though Monsanto had yet to achieve such approval in the United States or Canada. ACB similarly challenged an application by the ARC/CSIR (Council for Scientific and Industrial Research, a parastatal) for nutritionally improved sorghum on the argument that sorghum is an African heritage crop that risked both being privatized and having its enormous range of local varieties irreversibly depleted.[58] These challenges had the cumulative effect of paralyzing the regulatory apparatus for GMOs. In his 2005 annual report, the chair of the Executive Council of the GMO Act reported, "This is the first year that the anti-GMO movement has made active inputs into almost all . . . [GMO] applications submitted to the department. Although this is a good indication that public participation has increased in the regulation of GMOs, it is concerning to note that applications take a considerably longer time to be processed, which impacts negatively on the regulatory process and industries' perception of Government's ability to manage public participation."[59]

But if managing public participation was the government's concern, expanding public engagement with the issue was the activists' objective. As the GMO Act was being amended, activists pushed hard for expanded public notification regulations regarding the timing and location of GM field trials so that local citizens could respond. They also doggedly persisted with their campaign for comprehensive GMO labeling, with SAFeAGE lobbying heavily to have this requirement included in the Consumer Protection Act on the basis of consumers' *constitutional right* to choose what to eat. At the same time, ACB called for labeling under the "polluter pays" principle enshrined in environmental legislation in order to protect farmers from liability if something went wrong. In July 2007, ten years after the passage of the GMO Act, these activists managed to convince the Department of the Environment to hold public parliamentary hearings on the adequacy of labeling requirements for GMOs and organized a public campaign around these requirements. A year later, against heavy opposition from the Department of Agriculture, which sought to maintain regulation of GMOs under the GMO Act, and from the Department of Health, which adhered stringently to the principle of substantial equivalence, the Department of Trade and Industry's Portfolio Committee recommended that the act require products containing GM ingredients to be so labeled, though it modified the draft bill, removing language that would have required labeling of the "nature and extent" of such engineering, and left the management of labeling requirements up to the Department of Agriculture.

Although this outcome represented a very "soft" labeling regime, it nevertheless also represented a significant victory for a small group of activists whose campaign had started with a petition drive in July 1999 for food labeling when the Department of Health excluded the public from its deliberations on the issue. In the ensuing nine years, labeling requirements had been included and dropped from impending legislation twice and were opposed not only by key government ministries but also by Business Unity SA, which saw them as an expensive and technically difficult requirement. Anti-GM activists, however, hailed this development as a significant shift in the regulatory norms of agricultural biotechnology.

A Permanently Contentious Technology

In 2005, Ian Scoones commented on the conflict over agricultural biotechnology in South Africa: "A decade ago, GM crops were barely a concern in South Africa. The government, along with industry and a small cabal of scientists, set the terms. Today, this has all changed. A combination of high-profile court cases, ongoing demonstrations, a growing media profile and long-term engagement with legislators, bureaucrats and scientists has meant that the GM debate has been opened up to greater scrutiny."[60] The most important impact of this scrutiny was not a dramatic rise in public opinion against GMOs; indeed, public knowledge of GMOs remained scattered and spotty despite the polarized debate and the international networking of both sides in the conflict. Rather, the impact of this scrutiny lay in the ability of a small and deeply committed group of activists, drawing on local and transnational networks for information and support, to keep government regulatory agencies under pressure and to turn GM technology regulation into a zone of permanent contention.

To varying degrees, a similar pattern developed in other African countries where persistent opposition from vocal critics and the heightened scrutiny occasioned by the food aid crisis placed governments under powerful and contradictory pressures. Facing the scrutiny of activists and critical civil society organizations, African governments recognized the urgent need to erect effective regulatory frameworks. But that same scrutiny forced them to move very slowly and deliberately to construct biosafety frameworks and biotechnology policies, rather than simply adopting the permissive regulatory frameworks for which proponents of agricultural biotechnology had pushed and for which the South African GMO Act provided a model. Although those governments that had research capacity pushed ahead with transgenic research, the persistent uncertainty about the requirements of biosafety rules made it difficult for scientists to advance their projects lest they ultimately fail to meet the yet-to-be-determined standards.

While most African countries struggled throughout the millennial decade to build regulatory capacity and to pass biosafety legal frameworks compatible with the Cartagena Protocol, harmonization of biosafety frameworks across the region remained elusive, and the biotechnology industry found itself in an invidious position: unable to get its products into the market because of the absence of established and effective regulatory systems, but under intense surveillance from activist organizations should it try to exploit weak regulations. In August 2005, for instance, the Kenyan agriculture secretary ordered the destruction of all *Bt* maize crops undergoing field trials, because the environmental impact had not been fully assessed. In doing this, he pointed to "an emerging tendency by our scientists of yielding to pressure from international collaborators pushing to secure approvals for their research projects faster, side-stepping procedures."[61] In 2008, a public furor broke out in Kenya when activist groups discovered unapproved *Bt* maize in local crops planted with seed imported from South Africa. Meanwhile, activists across the continent continued to stress both the uncertainty associated with the environmental and health impacts of biotechnologies and the likelihood that they would enable foreign companies to extend control over local farming systems through the assertion of intellectual property rights. Consequently, as efforts to construct regulatory frameworks slowed down and became more contentious, debates about the potential impacts and risks associated with this technology became heightened, more wide-ranging, and more inclusive.

It would be wrong to suggest that the politics of contention engineered by activists closed the continent to GMOs. By the end of 2008, Burkina Faso had joined South Africa in approving the production of *Bt* cotton, and Egypt was moving rapidly in that direction. Nevertheless, the ground of biotechnology development had shifted substantially as a result of activists' obdurate opposition. In the first place, the closed world of agricultural technoscience had been cracked open, and the range of voices engaged in the governance of the technology had been broadened. In particular, both sides in the GM debate had set out to recruit "resource-poor farmers" to their position and, in doing so, had brought them into greater dialogue with governments over development challenges and choices. To be sure, the inclusion of farmers' voices on either side was sporadic and uneven, though activist organizations did broker the formation of a broad coalition of farmers' organizations in eastern and southern Africa to constitute an oppositional pressure group to reject GMOs from various states in the region.[62] But to the extent that debates over strategies for technology choices had become more public and inclusive, and farmers' interests and concerns had become part of the calculus of seed development, the centrality of transgenic technology in charting the way forward for African agriculture came into serious question.

A similar and perhaps more significant trend in new thinking about agricultural technology development and deployment can be discerned at the transnational level of policy formulation. In 2007, a dynamic new transnational initiative, the Alliance for a Green Revolution in Africa (AGRA), was established to rejuvenate smallholder farming in Africa in order to end widespread hunger and alleviate poverty. Heavily funded by the Bill and Melinda Gates Foundation and by the British Department for International Development (DfID), AGRA places a great deal of stock in science- and technology-based solutions for increasing agricultural productivity and reducing poverty in Africa. DfID's chief scientific adviser, Gordon Conway, had previously been president of the Rockefeller Foundation and was an enthusiastic advocate for transgenic technology. Nevertheless, much to the disappointment of biotech proponents, AGRA deliberately refused to foreground transgenic technology as a solution.[63] It took this position for three reasons. The first was that the seriousness of the agricultural and food crisis mandated faster and lower-cost strategies than transgenic research and development could provide. Investment in second-generation yield-increasing or nutrition-enhancing transgenic varieties had simply been insufficient to make them the backbone of a new green revolution.[64] Moreover, there was growing evidence that conventional plant breeding and nontransgenic biotechnology techniques could be very successful.[65] Indeed, this was where African public research institutions had their greatest capacity and had already established effective partnerships with international institutions such as CYMMIT (the

International Maize and Wheat Improvement Center) and CGIAR (the Consultative Group on International Agricultural Research).[66] The second reason was the increasingly influential argument that technology choices should be adapted to local regulatory capacity and choice, rather than adapting local regulatory systems to technologies. Indeed, one of AGRA's first initiatives was to launch a program for training African crop breeders to work "on local crops, in local environments [using] the power of applied plant breeding on African crops . . . to develop effective solutions to long-standing problems facing African farmers."[67]

The third reason that AGRA refused to foreground transgenic technology was that it was hindering solutions to problems because it had become so politicized after a decade of contention. Every decision regarding the development and deployment of this technology needed to be taken with full consideration of the likely responses of activists and the potential impact on public opinion. Willy-nilly, activists had forced the industry and regulators into a new and uncomfortable level of transparency. Moreover, the clear signal sent by AGRA was that if agricultural biotechnology was to play a key role in solving Africa's productivity and hunger problems, it must be approached in a different way.[68] By the end of the first decade of the new millennium, arguments that agricultural biotechnology can be effective only if it is substantially indigenized, with greater public support and resources, were gaining a currency in official and policymaking circles that had been absent in the first part of the decade.[69] Ironically, perhaps, the "slow race" of conventional breeding procedures may turn out to be the faster race to higher and more sustainable productivity.[70] If one recognizes that one way in which social movements have an impact is by *changing the normal ways of thinking about a problem,* then anti-biotech activists will have played no small role in bringing about this transformation.

A Different Future
for Biotechnology?

Social movements make a difference in a variety of ways. Some of their effects are immediately apparent, such as policy changes, while others take longer to demonstrate definitively, for example, cultural shifts. Some effects are short-lived and reversible, and others can change entire social or political systems. Measuring the efficacy or the "success" of a movement is consequently a complicated challenge. This is certainly true of the conflict over biotechnology.

If one takes the key desideratum of the movement—that is, the prohibition of transgenic technology in the production of crops and the manufacturing of food products—as one's measure, the movement certainly has not succeeded. As we showed in preceding chapters, even as the movement was surprisingly effective in Europe, activists failed to attain their immediate goals in other places. In the United States, the industry swept aside a vigorous campaign to stop the deployment of recombinant bovine growth hormone and crushed GM food-labeling campaigns. In parts of Africa where the importation of GMOs was not yet legal, GM seed crept inexorably into the seed supply.[1] Global plantings of the core GM crops have expanded steadily since they were introduced in 1996, and after a significant shake-up in the late 1990s and early 2000s, the agricultural biotechnology companies have again come to enjoy considerable success in the market. The sense of industry panic with which we began this book has greatly receded.

Changes in conditions surrounding the debate over agricultural biotechnology played an important role in facilitating its resurgence. Late in the first decade of this century, rapidly rising oil prices sparked the interest of many governments in expanding the industrial production of agricultural crops for processing into biofuels. Also, world food prices hit an all-time high in the spring of 2008, reflecting the dramatic deepening of a world food crisis that had been in the making for quite some time.[2] The biotechnology industry and its allies in the scientific and policymaking communities were quick to capitalize on these developments. Virtually as soon as the specter of a world food crisis appeared in the media, they began arguing that it was time to stop resisting GM technologies and to move beyond "irrational" fears of biotechnology, since it had already been used successfully for ten years. Seizing the opportunity to push forward their agenda, they sharpened their calls for governments to relax regulatory constraints on biotechnology and to let scientists and industry proceed with the job of raising agricultural productivity through its use. Just as they had during the 2001–2002 central African famine, these proponents blamed a small group of well-fed, elite activists for obstructing a solution to the world hunger problem, loudly chiding them for being cavalier about the suffering of the poor.

In this new context, anti-biotech activists were much harder pressed to defend their criticisms of a technology that at least in theory held the promise of augmenting agricultural productivity and addressing the new challenges facing agriculture, including the higher demand for food, the quest for non-fossil-based fuels, and climate change. Indeed, in light of the notable lack of any environmental or public health disasters associated with transgenic technologies, many farmers' apparent enthusiasm for them, and the labor reductions that these technologies had shown in some cases, the activists' uncompromising position that agricultural biotechnology had nothing to offer became increasingly difficult to sustain. By the end of 2008, it was impossible to escape the conclusion that agricultural genetic engineering is here to stay and will continue to be an important component of global agriculture.

Nevertheless, the world of agricultural biotechnology today is not the world that the industry envisioned. In important respects, Bruce Skillings's observation at the FDLI conference that "the future ain't what it used to be" has been manifested: the industry and the technology have not followed the anticipated path of growth and development. If the industry had had its druthers, no one would have questioned the wisdom of using these technologies or worried seriously about their impact on the environment. Genetic engineering would have been more widely adopted across different kinds of crops and in many more countries. The industry would have been regulated but not in a manner that would have interrupted the smooth flow of products to markets across the globe.

The technology would have been recognized as novel for the purpose of claiming intellectual property rights but as nothing new when it came to the need to evaluate its potential effects on human health or the environment. The insertion of patented genes would have been recognized as the principal solution to the problems of agricultural productivity, pest management, and global food production.

Instead, however, the expectations and hopes for the gene revolution have diminished significantly by the end of this century's first decade. GM crops and products have been rejected in crucial markets and by certain publics. Plans to market some key genetically modified crops, such as GM wheat and potatoes, have been shelved or considerably delayed. In some countries, such as India and Brazil, the deployment of GM crops has escaped the control of the industry, as farmers and local seed producers have taken seed reproduction into their own hands.[3] New systems of biotechnology regulation and governance have been established at the national and international levels, broadening the range of actors involved in decision making and giving governments greater say in the adoption and promotion of new technologies. The issue of novelty has become a double-edged sword: while firms have been successful in convincing courts, patent offices, and the public that these technologies are revolutionary, that very claim has made the public nervous about their potential effects on human health and the environment. Certain of the questions and concerns raised by anti-GM activists have come to be routinely and procedurally included in decision-making processes related to biotechnology in the spheres of both regulatory policy and firm investment strategy. The most obvious example is the inclusion of environmental risk criteria in government regulatory policy, although consumer right-to-know issues (especially labeling) have also become important considerations. In addition, the biotechnology industry now makes its decisions about whether to invest in particular technologies with the long-term costs of regulation and possible social resistance in mind. In effect, the cognitive frameworks and institutional environments in which agricultural technology development takes place have been substantially altered.

Indeed, agricultural biotechnology has become intensely politicized over the past decade and its chief developers (large multinational corporations) and its chief supporters (many governments) have been subjected to intense public scrutiny. The arenas in which technology decisions are made have become more crowded, as many new actors have become involved in negotiating the role of biotechnology development, especially in the global South. These actors range from local farmers' organizations to public research institutions (including university-based plant biology departments) to international aid agencies, charitable organizations, and private foundations. Many of these groups are committed to raising agricultural productivity, alleviating hunger and poverty, and helping poor farmers. Some, such as the ISAAA (the International Service for the Acquisition of Agribiotech Applications), the African Agricultural Technology Foundation, and the U.S. Agency for International Development, regard modern biotechnology as the only realistic means of achieving these goals. But other groups, such as Oxfam and the Alliance for a Green Revolution in Africa (AGRA), are sensitive to the inability of this technology—at least in the way it has been developed in the last twenty years—to deliver on its promise. Indeed, a widening circle of biotech proponents appear to be reaching this conclusion. As an August 2008 editorial in the pro-biotechnology journal *Nature Biotechnology* put it, the claim that transgenic technologies offer the most promising means of healing, feeding, and fueling the world is "an outrageous act of faith bordering on the religious."[4] Reflecting this more considered view, AGRA's chairman, the former UN secretary general Kofi Annan, explicitly downplayed the significance of GMOs as the key to African agricultural development, because the technology was so politicized that it might actually impede effective efforts to resolve the continent-wide crisis of agricultural productivity. Likewise, some funding agencies, such as the Rockefeller Foundation, have partly shifted their focus away from GM technology with the recognition that the high hopes of the technology simply cannot be attained. In short, at the global level, the climate in which the possibilities of genetic engineering can be imagined, let alone pursued, has changed markedly.

Similar shifts have occurred at the regional and local levels, where groups explicitly concerned with asserting the interests of farmers in determining technology choices and development strategies have come to play an increasingly active role. Most of these are civil society organizations and farmers' associations. They have added substantially to the plethora of *nonactivist* actors that have become engaged in the issue and are adopting more critical and nuanced perspectives on the potentials and prospects of transgenic technologies. These actors understand that the challenges facing farmers and the poor have complex socioeconomic, political, and institutional roots and therefore require a variety of strategies. To them, technology is important, but the role it can play is only part of what's needed. And its potential will surely be limited if it is not accompanied by other political, economic, and institutional changes.[5] They see the need to push the debate past the stultifying binary oppositions of

GM or non-GM futures and to direct research and resources to a wider range of alternatives.

In effect, then, fifteen years after their initial commercial deployment, agricultural GMOs have certainly spread around the globe. But they have not done so in the way or with the ease that proponents expected they would. Nor have they done so without becoming one of the most highly contested technologies of our time. In early 2010, the box of potential solutions to the challenges of agricultural productivity and sustainable development in the twenty-first century looks far more open than it did ten years ago. The criteria on which these solutions are to be judged have expanded significantly. And the range of voices debating them has become much wider. The course of this technology has been altered significantly, and its future, once so clearly envisioned by its proponents, is far less assured.

Assessing Efficacy: How Anti-Biotech Activism Mattered

Our objective in this book has been to explain the role of the anti-biotechnology movement in producing the outcomes described above. Yet the contradictory trends we have just discussed suggest that this is not a straightforward analytical task. The causal impacts of activism, either on movement targets or on society in general, are notoriously difficult to isolate and measure. These effects are often variegated or indirect. Much depends, therefore, on how one interprets and measures movement impacts. The easiest way to assert the efficacy of a social movement is to show a change in public policy related to the demands of activists. But even this outcome can be misleading. As several analysts have pointed out, the formation of public policy is shaped by a variety of factors, including public opinion, the degree of electoral competitiveness, the strength of special-interest organizations (including social movements), and the authority of experts.[6] Certainly, the process of change, including policy change, is not only defined by the actions of movement activists or by their contentious interactions with adversaries. Activists are only one set of actors in a complex political field of interactions among a range of social and political actors, not all of whom are directly engaged in contention. How social movements influence social and political conditions is thus contingent on a variety of factors over which they have little control. Moreover, the most important impacts of a social movement may not be directly related to their specific objectives. As Marco Giugni and colleagues have noted, movement actions can affect politics, culture, and policies in ways that are both unexpected and unintended and that may, in fact, be undesirable in terms of movement goals.[7] This is particularly true in transnational conflicts, in which the political opportunity space, the actors themselves, and contentious interactions are complex and multilayered in nature.

These complexities press us to question not only to what extent the "success" or "failure" of social movements is ascribable to the movements' own efforts but also, indeed, whether success and failure are the appropriate measures of a given movement's influence. A more fruitful approach is to ask: What difference do social movements make to social and political change? This "how movements matter" approach bears several implications for studying a social movement's influence.[8] In the first place, it suggests that we need to look beyond what social movements *intend to achieve* through their strategic actions and focus instead on the processes of change that social movements *actually set in motion* as they pursue their objectives. Such an approach explicitly recognizes that the impacts of movements are typically bound up with a complex series of events involving the actions and reactions of a multitude of stakeholders and other actors. This means that it is difficult to isolate the effect of a movement's actions or to argue that a particular action caused a given outcome. This is especially true in multilevel transnational movements, in which influences can "bounce" reflexively across space and level. Searching for a parsimonious explanation may misrepresent the actual impact of specific actors.

The "how movements matter" approach also indicates that movement impacts can take a variety of forms and can manifest themselves at multiple levels. Some impacts are specific and direct; for instance, a movement might force a government to make a particular policy change or to change the operating procedures of institutions, for example, by requiring companies to conduct environmental impact assessments of their proposed waste disposal projects. Other impacts are broader and may become apparent only in the long term. For example, as Sylvia Tesh has shown, the most significant impact of the U.S. environmental movement has been a long-term cultural change in public attitudes toward the environment.[9] Some processes of social and political change are not necessarily driven strategically by movements but would not have happened (at least in the form that they did) in the absence of the movement. In order for us to take account of such processes, it is useful to ask the counterfactual question, Would the situation, or process of change, have been much different had this movement not emerged? By moving away from questions about a movement's "success" and placing the movement's influence within a broader political field of actors, institutions, and historical contingencies, this question allows one to identify more fully the range of consequences associated with mobilization.

The anti-biotech movement is clearly a movement that should be assessed not in terms of its success but in terms of *how it mattered*. The most profound impact of anti-biotech activism was to establish the distinction between genetically modified organisms and non–genetically modified organisms as the defining social and technical fact about the technology.[10] This binary distinction became the discursive fulcrum on which opponents and proponents constructed contending discourses concerning the nature and meaning of bioengineering. In a masterpiece of evocation, Greenpeace International coined the term *Frankenfoods* to define all GMOs as unnatural, uncontrollable, and ultimately unpredictable (and for some an immoral adventure in "playing god with nature"). This "Frankenframe" provided rich source material for the costumed street theater that is a hallmark of contemporary mass movement activities. More important, however, the binary opposition between GMO and non-GMO had a classificatory effect that allowed activists to present genetic engineering as an *inherently* unacceptable approach to the social and technical challenges of agricultural productivity and food supply, because it harbored incalculable risks of irrevocable disaster. In Europe, as we have seen, activists were able to link the term *genetically modified organisms* (or its acronym, GMO) to a loss of consumers' choices of what they might eat. In the global South, activists were able to connect the term to profound, if unspecific, threats to farmers of losing control of their seed supply, farming systems, and possibly their land. In India, activists worked hard in the first decade of this century to link the planting of GM cotton to the increase in farmers' suicides. In short, the term became ineluctably loaded with normative and political meanings that erased any possibility of asserting its scientific neutrality.[11]

This discursive framing turned out to be enormously powerful. In Europe, activists' framing of genetic engineering as inherently unsafe and untested led many consumers to reject genetically engineered products in the marketplace. Given the global nature of agricultural markets, these decisions reverberated in producer countries, where the potential loss of markets posed a dire threat to agricultural exports. Many governments became understandably nervous about adopting a technology that was widely regarded as risky, and they moved very cautiously when it came to allowing GMOs into their agricultural systems. Particularly in Africa, where so many livelihoods depend upon agriculture and where economies are extraordinarily fragile and regulatory capacity is weak, governments were fearful of the potentially devastating effects of this technology if something went wrong.

By raising consumer consciousness in the advanced industrialized countries and increasing the risk of poor market returns on GM products, activists made biotechnology companies more reticent to pursue what were already expensive and risky research projects. This reticence was compounded by the escalating costs of product development, attributable in part to the anti-biotech movement's actions at the regulatory level.[12] As activists assumed a watchdog role over GMOs and posed a continuous stream of challenges to national regulatory authorities, individual products became subject to longer regulatory delays, which made bringing them to market more expensive for companies. In some cases, a product failed to make it through a given regulatory system at all, which represented a serious setback for the companies involved. Increased regulatory oversight and tighter regulations in many countries raised the uncertainty surrounding investments in biotechnology and, hence, the financial risk associated with new R&D projects. Not only did this lead agricultural biotechnology companies to revise their own assessments about whether the pursuit of particular GM applications was worthwhile, but it also caused Wall Street to downgrade its evaluation of the likely profitability of these firms. Together, these processes served to narrow the set of technologies in which the industry was willing to invest.

The binary framing of the technology worked differently in the global South, where many people still make their living from agriculture, and farmers do not expect to buy all their seed, especially from foreign corporations. Since the research and development costs of genetic engineering are high, the foreign companies producing GM seed sought to mitigate the economic risk of developing particular crops by picking crops from which they were sure to see a timely return and by controlling the technology through implementation of patent rights, licensing fees, and contracts that prohibit farmers from saving, replanting, and breeding seed. While proponents of the technology invoked the GM/non-GM distinction to stress its broad potential to "heal, fuel, and feed the world," critics invoked that same distinction to represent the technology's power to entrench corporate control of farmers through the assertion of their intellectual property rights in seeds.

If we return to the counterfactual question that we posed earlier—*Would the current situation be the same had this movement not occurred?*—we see now that the answer is clearly no. By problematizing the value of agricultural biotechnology and questioning its promise as a social good, activists prevented the technology's proponents from deploying it in their own ways and on their own terms. In the absence of mobilization against biotechnology, no meaningful social distinction between GM and non-GM products would have developed. The industry would have enjoyed a higher return on its investments. GM seeds would have spread much more smoothly, quickly, and extensively across the globe. The technology would have been regulated very differently and more permissively in many countries. And GM crops would far more likely have been accepted as the obvious solution to global hunger, rising food prices, and low agricultural productivity in poor countries, rather than being viewed as one strategy among several for dealing with such problems. In short, the "David" of the anti-biotechnology movement had seriously disrupted the course of the "Goliath" of the biotechnology industry. In this sense, the movement mattered immensely in charting the course of this technology.

Explaining Efficacy: Opportunities, Lifeworlds, and the Dynamics of Contention

As we have argued in these chapters, the power of the anti-biotech movement resided not in its numbers (for this was clearly not a mass movement) but in its ability to exploit the vulnerabilities of a globally organized industry and in its members' fierce and unrelenting commitment to the issue, which arose out of what we have called the movement's lifeworld.

The movement's ability to identify and exploit the vulnerabilities of the agricultural biotechnology industry successfully derived largely out of the industry's economic organization, particularly its global character, although industry culture also played an important role. The fact that the agricultural biotechnology industry was situated at the upstream end of a global commodity chain over which other industries (namely, the supermarket and processing industries) and national governments exercised considerable control rendered it highly vulnerable to the actions of these other chain actors. As we documented in chapter 4, the relations of dependency that characterized this particular commodity chain offered activists the possibility of intervening at several alternative nodal points once they discovered they had little direct leverage over agricultural biotechnology companies themselves.

Emergent national and supranational regulatory systems provided activists with additional points of entry and influence. Not only did different organizations and local movements play important roles in shaping some of these systems, but they also helped to ensure that more stringent national regulatory systems would be put into place, particularly in the global South. Again, the global nature of the agricultural biotechnology industry rendered these firms vulnerable to social activism. Because companies were required to obtain government approval from every country to which they sought to sell or import GM food, seed, or feed, the opportunities for activist intervention multiplied dramatically. Activists were keenly aware of these opportunities and used them to their advantage. Together, these multiple dimensions of "industry globalism" augmented the movement's power beyond that which it would have wielded had these struggles been purely national. In effect, the global reach of the agricultural biotechnology industry was also its Achilles' heel.

A final source of the agricultural biotechnology industry's vulnerability arose out of its cultural characteristics and the historical experiences of the people in the countries where these firms sought to introduce their technologies. Just as activists were driven by the way they saw the world, industry actors were also driven by the worldviews and the behaviors that were considered normal within their lifeworlds. Dominated by a powerful and hyperaggressive U.S. firm, whose executives were convinced of the benefits of biotechnology and whose economic investment in the technology compelled it to go to the limit in marketing it, the agricultural biotechnology industry went charging ahead as it sought to introduce its products overseas. Monsanto's aggressive behavior and the overblown claims it and the rest of the industry made about biotechnology left the industry wide open to the criticisms that it was glossing over the real risks of the technology and that it did not respect various publics' rights to decide what kinds of technologies they wanted. Recognizing that Monsanto's behavior would deeply offend public sensibilities, activists were able to make the company into an icon of corporate greed and imperialism.

The Significance of Lifeworlds

In this book, we have used the concept of lifeworlds to help us understand why industry and activists engaged in a thirty-year struggle over agricultural biotechnology. We have argued that the contending lifeworlds of these two protagonists have defined the terms of debate about the value of transgenic technology as a public good. For activists, the cognitive perceptions of the technology associated with their lifeworld were buttressed by the persistent *uncertainties* embodied in the technology, which manifested themselves in the steady, if uneven, trickle of evidence from various parts of the world that GM technologies could not be contained, were becoming increasingly ineffective against weeds and pests, did not consistently improve yields, and were ineffective in reducing poverty.[13] Part of the activists' mode of operation was to doggedly search for these uncertainties; each discovery helped to confirm their suspicion of the technology. In this way, their lifeworld also reinforced these actors' sense that their cognitive perceptions of the technology—and of its developers' motives and modus operandi—were correct.

Inextricably linked to these cognitive perceptions was a set of normative concerns about the technology that helped explain these activists' unwavering commitment to the cause. These normative concerns—worries about the privatization and commodification of the "basic building blocks of life," about irreversible environmental degradation, about the impact on farmers in the global South, about the use of science for private gain rather than public good—derived from a sharp awareness of the political–economic and institutional context in which these technologies were being developed. Indeed, for these actors the technology could not be interpreted outside this context, which comprised the "neoliberal revolution" of the 1970s and 1980s that profoundly altered the roles and behavior of the public and private sectors. This revolution included a set of legal developments that extended private property rights into uncharted territories, including the realm of genes, gene sequences, and whole organisms. Recognizing the logic of this institutional environment for driving biotechnology firms' behavior, critics were led to see greed and economic rapaciousness wherever these companies were to be found. This instilled in them a profound distrust of life science firms and their technologies and fueled a deep sense of outrage about what these companies were doing.

For their part, the adherence of biotechnology proponents to a vision of science- and technology-driven progress generated an equally strong sense of outrage about activists' behavior. Unlike the activists, proponents did not understand the institutional context as being an essential component of the technology. For them, the utility and value of any particular technology was inherent in the technology itself, as an object existing apart from society. This conviction underpinned their passionate commitment to transgenic technologies as *good technologies,* either because they could generate profits for their companies or because they offered solutions to some of humankind's greatest challenges, such as feeding a growing world population. Industry leaders believed that the market would serve as an objective arbiter of value and should be left largely unfettered. Scientists were committed to producing innovative knowledge in a context that rewarded ingenuity and hard work. The contending lifeworlds of industry and activists drove the controversy over biotechnology, pushing each side into uncompromising and mutually unintelligible positions.

By seeing these *contending lifeworlds* as possessing their own coherence and logic, we can more easily comprehend the behavior of movements and their adversaries as they sought to pursue their goals and developed a strategic plan of action toward the other. But it is important to bear in mind that activist and industry lifeworlds were not symmetrical in relation to each other. For those who were part of the industry lifeworld, the ideas, beliefs, and commonsense ways of acting that informed their lifeworld were largely there for the taking; that is, they drew on widely shared ideas and norms in society that were part of taken-for-granted and unspoken "realities." For those who were part of the activist lifeworld, however, this was not the case. Because their lifeworld was inherently *oppositional* and therefore not shared by most members of the larger society, they could not draw heavily on commonsense discourses, worldviews, and norms that already existed in society. As we saw in chapter 3, these activists had to construct the critical analytical and normative frameworks of their lifeworld consciously through serious and sustained intellectual work. Thus, the terms of their lifeworld were largely "spoken," and the sense of community on which they drew was voluntary, intentional, and conscious.

In short, the activist lifeworld was inherently purposive: the combination of ideas, values, and community that constituted it not only disposed activists to particular ways of thinking and seeing the world but also underpinned their sense of commitment, urgency, and passion. Through their analysis of the developments occurring in the biological sciences, their critiques of the social relations surrounding the new biotechnologies, and their challenges to the emerging legal frameworks governing these technologies' ownership and use, these biotechnology critics forged an oppositional ideology and alternative discourse upon which a movement could be built. Indeed, without the intellectual work conducted by this critical community of activists in the 1970s and 1980s, an anti–genetic engineering movement would not have developed in the 1990s.

As the struggle over agricultural biotechnology demonstrates, a consideration of social actors' lifeworlds helps us to develop a fuller account of contentious politics and its dynamics. Most important, this consideration emphasizes the idea that both activists and their adversaries are *cultural actors*.[14] Their actions and their interactions reflect particular cultural logics and assumptions. Their ways of understanding and of being in the world—the assumptions they bring to particular situations, how they evaluate and make use of information, the informational sources they value, trust, and depend upon, and the way they read the actions of others (and act in response)—all reflect the influence of their lifeworlds. The lifeworld can be a source of considerable power to one protagonist or the other. For instance, in the case of the anti-biotech movement, the lifeworld provided a strong sense of community among group members, as well as a shared understanding of the world that gave them a powerful sense of confidence that their understanding was *right*. These sensibilities were what enabled activists to cling tenaciously to their cause, even under highly inauspicious conditions for political organizing, and thus contributed to the persistence of the movement through time. Ultimately, the simple refusal of the movement to go away played a significant role in determining activists' impact on the trajectory of the technology.

But the lifeworld can also seriously limit the worlds that actors can imagine and can thus constrain the kinds of strategies they are willing to adopt. Monsanto's decision to introduce GMOs into Europe in a way that was consistent with the *company's* cultural rationality (that is, introducing them into the marketplace unlabeled and unannounced) is an excellent example. This cultural misreading created a critical opening for the European anti-biotech movement to attack. Thus, an understanding of different actors' lifeworlds can give us a handle on the cultural rationalities that shape their readings and behaviors and thus illuminate why they might act in ways that are simultaneously *counterproductive* and yet *entirely sensible*.

The dynamic interaction between adversaries' lifeworlds significantly affects the nature of a struggle, in part because it influences the degree and kinds of compromises that adversaries can reach at any particular moment. In other words, the lifeworlds of movements and their adversaries define both the potential spaces of agreement or concession and the positions that are absolutely nonnegotiable at any particular moment. In the biotech struggle, for instance, intellectual property rights represented one of the fundamental oppositions at the core of the conflict. Industry actors believed implicitly that property protection was essential for innovation. Activists, by contrast, saw intellectual property protection as inimical to promoting the public good and as a violation of one of their most deeply held values, that is, that life itself should not be subject to private property relations. This was thus an issue on which neither would, or could, compromise. Unless and until the meaning of the technology could be disarticulated from private property relations, this conflict could not be resolved. A lesson for social movement theorists is that one can learn much by seeking to understand adversaries' lifeworlds and what is negotiable and nonnegotiable within them.

It is important, finally, to recognize that the role of lifeworlds in shaping social struggles changes across time as the terrains and terms of contention change. Such shifts result in part from the dynamics of contention itself, but they may also result from exogenous changes in the environment, such as a significant change in political alignments or a collapse in world market prices for a strategic commodity. The ability of adversaries to adjust to such changes will inevitably be conditioned by their respective lifeworlds. Their lifeworlds will also shape the role they continue to play in producing social change. For instance, a rigid adherence to a particular worldview can delegitimize a group or render its discourses less persuasive or both. This is perhaps the situation that confronts the anti-biotechnology movement today. Although activists' steadfast commitment to certain precepts served the movement well for most of its existence, the political and economic climate surrounding the technology changed after 2005, and the scientific data amassed on particular applications of the technology have been augmented. As a result, what was once a source of strength for the movement—namely, a powerful commitment to particular ideas and claims—has become a source of weakness, as the movement has rendered itself incapable of seeing *any* conditions under which agricultural biotechnology might be beneficial.[15] If one of the movement's signal achievements has been to help expand the range of imaginable agricultural futures, it may, in so doing, have limited its own ability to contribute to the construction of such futures. Much will depend on how both sets of adversaries read and respond to new opportunities.

ACKNOWLEDGMENTS

As authors working at different institutions in different cities, we earned a great many debts of gratitude in the course of researching and writing this book. First and foremost, we thank the many activists, industry officials, scientists, and other academics who allowed us to interview them about their work on agricultural biotechnology. We are grateful to them all for sharing their experiences, reflections, and insights, enriching our understanding of the biotechnology controversy.

Research projects have their own intellectual histories that draw on preceding ideas and the work of others. While it is impossible to mention all such debts, we acknowledge two particularly significant influences on this project: one is the work of Jack Kloppenburg Jr., whose analysis of the commodification of the seed profoundly shaped our understanding of capitalist agriculture, and the other is Doug McAdam, a great student and theorist of social movements, who prodded us to deepen our understanding of industry as a contentious actor in the controversy over biotechnology.

We began working on this project when we were both teaching in the Department of Sociology at the University of Illinois. There we received encouragement and comments from many colleagues and friends, especially Zsuzsa Gille, Michael Goldman, Zine Magubane, Anna-Maria Marshall, Faranak Miraftab, Tom Bassett, Kathryn Oberdeck, Ken Salo, and John Lie. Dana Rabin, Nancy Abelmann, and Ingrid Melief were such a crucial part of Rachel's life that she cannot imagine Champaign–Urbana without them, and Faranak Miraftab was most important of all, housing Rachel on her frequent visits back to town, regaling her with stories of life, love, and kids, and feeding her delicious Iranian food.

Since then, our work has been nourished by people in several far-flung locations: our new workplaces in Bloomington, Illinois, and Minneapolis, Minnesota; the Agrarian Studies Program at Yale University; and the School of Development Studies at the University of KwaZulu–Natal. At the University of Minnesota, Rachel had the good fortune to participate in a writing group with Jennifer Pierce, Lisa Park, and Teresa Gowan: all made significant contributions to this project, and we thank them. Teresa has been here from the start of Rachel's life in Minneapolis to offer a constant source of friendship, inspiration, and unparalleled intellectual engagement. Ron Aminzade has also been a great friend, department chair, and colleague, and we thank him for his support, intellectual engagement, and sense of humor. Thanks, too, to Evelyn Davidheiser and Klaas Van der Sanden at the University of Minnesota's Institute for Global Studies; to Mary Drew, Ann Miller, and Chris Uggen in the Department of Sociology for creating a supportive work environment; and to Rachel's agrifood reading group (especially Valentine Cadieux, Rachel Slocum, Tracey Deutsch, Jerry Shannon, Joaquin Contreras, and Ursula Lang) for their comments on our introductory chapter. Jerry Shannon focused his skilled editorial eye on several other chapters of the manuscript and deserves an extra dose of thanks. Anne Kapuscinski has also been a valuable interlocutor and friend since Rachel met her in 1997.

At Illinois Wesleyan University, William was fortunate to have a group of colleagues who brought their own intellectual interests in food, health, and environmental sustainability to bear on our work and pushed us to clarify our arguments. We especially thank Irv Epstein, Rebecca Gearhart, Abigail Jahiel, and Chuck Springwood. Regina Linsalata, administrator of the International Studies Program, freed up precious work time by fielding numerous interruptions. Patrick Beary and Mike Feeney offered excellent research assistance, hunting down obscure bits of data with sustained good humor.

Throughout this research, William benefited from the warm hospitality of the School of Development Studies at the University of KwaZulu–Natal in South Africa. Special thanks are due the head of school, Vishnu Padayachee, for generously accommodating William's repeated visits and arranging his status as research associate. Imraan Valodia made his office (and remarkable library) available, and Glen Robbins, Caroline Skinner, and Dori Posel suggested new lines of inquiry and sources of information. Harald Witt kindly shared his own research on small-scale cotton farming in the Makhathini Flats region. It is hard to imagine a more supportive and stimulating environment in which to complete a project.

Several others gave us helpful feedback at various stages. Ron Herring, Joost Jongerden, Guido Ruivenkamp, and Wietse Vroom provided sharp if friendly critiques of early presentations that helped us to refine our idea of lifeworlds. Tom Bassett swapped ideas on African biotechnology during long runs through the cornfields of central Illinois; his friendship throughout this project has been invaluable. We also thank Daniel Kleinman, Scott Frickel, and Jim Saliba for their useful comments on various chapters. Dr. Robert Archer provided a good critical read of our chapter on the industry and usefully helped us clarify certain points. Rachel's sister-in-law Kate Dunnigan is not only a dear friend but a terrific historian and editor as well; while in the midst of her own travails, she generously focused her acute mind on the manuscript, improving it immensely. One of Rachel's graduate-school friends, Susan Pastor, gave several chapters a thorough read and edit, for which we are grateful. Sue also lent us her personal archive on the anti-rBGH movement in Madison, Wisconsin, in the late 1980s.

Long research projects cannot be accomplished without sustained institutional support. For research funding, we thank the Research Board at the University of Illinois; the faculty development program at Illinois Wesleyan University; and the College of Liberal Arts, Institute for Global Studies, and Department of Sociology at the University of Minnesota. Rachel acknowledges fellowships from the Program in Agrarian Studies at Yale University and the Institute of Advanced Study at the University of Minnesota: both offered her much-needed time to think and write in an atmosphere of rich intellectual engagement. We thank Vicki Bierman for carefully transcribing our interviews.

Andrew Szasz read the entire manuscript and offered extremely valuable advice for strengthening and tightening our argument. At the University of Minnesota Press, Jason Weidemann shepherded the manuscript through the publication process with great patience.

Finally, we thank our families, without whom this book could not have been written. Rachel's brothers, Josh and Paul, and her sister-in-law Kate kept her laughing during her visits to Rhode Island, and her sister-in-law Barbara welcomed her into her home for good meals, wonderful company, and (yes, we can't live without it) Internet access. Rachel's parents, Bertha and Bernard Schurman, were always there with love, warmth, and wisdom. Although neither lived to see this book come to fruition, she knows that they are here in spirit, reveling in its completion. Bertha never ceased to inspire her family and friends with her invincible energy and spirit; she stands as an example of how to live life to the fullest. In

Durban, William's mother (herself an indomitable figure) remained resolutely interested in the project while opening her home to periodic floods of family, friends, and visitors. Her legendary hospitality deeply enriched our visits.

We are remarkably lucky to have spouses who are willing to put up with our long interactions on the telephone and Internet, as well as our frequent invasions of each other's homes for intense work sessions. Michael Goldman and Kathy Oberdeck have both been willing to pick up extra household responsibilities so that we could work together, and we look forward to supporting them in their own research endeavors.

We thank our children, Nadia and Eli, and Fiona and Cara, for bringing so much joy into our lives and for tolerating our bouts of distractedness and periodic travels away from home. The future of food is also their future, and we hope that the struggles we recount in this book will create new possibilities. We dedicate the book to them.

Data Sources

The data on which this book is based come from a wide range of sources. Between 2000 and 2007, we conducted more than eighty in-depth interviews with activists, scientists, corporate executives, and others. About three-quarters of these interviews took place in person, and the rest were conducted by telephone. With a handful of exceptions, we taped and transcribed every interview. These interviews took place in the United States, Europe, South Africa, and India.

Another important data source comprises the extensive press coverage the GMO issue attracted around the world. Newspapers and magazines were the most common source of this coverage, but we also relied heavily on various specialty or "trade" journals for information, including *Chemical and Engineering News, Science, Nature Biotechnology, Delta Farm Press, GeneWatch,* and the *Grocer.* A large proportion of the media coverage surrounding agricultural biotechnology is systematically documented by NGOs around the world and is made available daily on the Internet and various Listservs dedicated to agricultural biotechnology. Two Listservs we diligently followed were the biotech activist Listserv run by the Institute for Agriculture and Trade Policy, in Minneapolis, Minnesota, and GENET, the news and information service maintained by the European NGO Network on Genetic Engineering. We supplemented these sources with information from activist, industry, and government Web sites, which included annual reports and other reports, policy and position statements, news articles, and press releases.

We gleaned information from a number of secondary sources as well. Among them are Daniel Charles's book *Lords of the Harvest,* Belinda Martineau's insider analysis of the biotech startup Calgene,[1] Derrick

Purdue's early study of the European anti-biotech movement,[2] and the excellent work of Les Levidow and his colleagues on biotechnology in Europe.[3]

Finally, we engaged in participant observation at six activist and industry conferences, held in Washington, D.C. (October 2001); Amherst, Massachusetts (November 2002); St. Louis, Missouri (two held back-to-back, June 2003); Urbana, Illinois (April 2004); and Berlin, Germany (January 2005). Three of these were organized by activists, two by industry, and one by a university. These conferences not only provided an important opportunity to listen, observe, and talk formally and informally to a wide range of people involved with the biotechnology issue, but were also generative of a number of our insights and theoretical ideas.

Introduction

1. These included the Biotechnology Industry Organization, the Grocery Manufacturers of America, and the National Food Processors Association.

2. A sampling of the participant list reveals the variety of organizations that attended the conference. From industry, representatives from Monsanto, DuPont, Syngenta, Cargill, Pioneer Hi-Bred, H. J. Heinz, the Campbell Soup Company, Novigen Sciences, General Mills, United Sugars Corporation, and Integrated Coffee Technologies were all present; from the media, reporters hailed from *Food Chemical News, The Economist,* and *Pharmaceutical Executive Magazine;* from the U.S. government, officials came from the Departments of Agriculture and State, the Food and Drug Administration, the Environmental Protection Agency, and the Agency for International Development; and from the finance sector, analysts from Fidelity and several other firms attended. The list of participants also included lawyers working in the food and drug sector, representatives from trade associations, and a few academics.

3. Author's notes on Rudenko presentation, FDLI Conference, October 29, 2001.

4. Gaskell et al. 2006.

5. Author's notes on presentation by Steve Daugherty, director of international affairs at Pioneer Hi-Bred, FDLI Conference, October 29, 2001.

6. Biodiversity Action Thailand 2000.

7. Roosevelt 2000.

8. The United States has consistently been the leader in GM plantings. The other two countries that had taken up the technology enthusiastically by the time of the FDLI conference were Argentina and Canada. Since then, Brazil, India (cotton only), China, and South Africa have joined the list of major adopters (James 2008).

9. See Borlaug and Dowsell 2001.

10. The Green Revolution refers to a set of agricultural technologies developed after World War II that greatly increased agricultural productivity, particularly in Asia and parts of Latin America. The centerpiece of these new technologies comprised new, high-yielding varieties of rice, maize, and wheat, which had to be used in conjunction with synthetic nitrogen fertilizer, irrigation, and pesticides to achieve their full potential.

11. The framing literature has usefully drawn attention to the significance of interpretative processes in grievance formation; see Snow and Benford 1988; and Snow et al. 1986.

12. We take the phrase "shared mental worlds" from James Jasper, on whose work we draw (Jasper 1997).

13. Prior to Schutz's and Habermas's development of the concept, this term was used by a group of early twentieth-century German philosophers who developed an experiential theory of knowledge known as phenomenology. Central to phenomenology is the idea that people come to know, understand, and attribute meaning to (natural and social) phenomena through their everyday "lifeworlds." Schutz and Habermas engaged directly with these philosophers but took the concept into new terrains. Dilthey (1833–1911), Husserl (1859–1938), and Heidegger (1889–1976) were among the most prominent theorists of phenomenology.

14. Habermas 1984, 124–35; Schutz and Luckmann 1973, 4.

15. In Habermas's words, "The lifeworld appears as a reservoir of taken-for-granteds, of unshaken convictions that participants in communication draw upon in cooperative processes of interpretation" (1984, 124).

16. Habermas discusses the lifeworld concept in his magnum opus, *The Theory of Communicative Action*. The bulk of this book is dedicated to developing a broad theory of human language and communication, although ultimately Habermas addresses the question of how societies evolve from premodern to modern, positing a process by which the bureaucratic rationality of modern society gradually imposes itself on and transforms (or *colonizes*) the "communicative rationality" of premodern societies.

17. Schutz views the lifeworld as inherited. Habermas also views the lifeworld as inherited but recognizes its active construction by humans through the use of language. Consistent with our more grounded interpretation, we view lifeworlds as being actively constructed in particular times and places by particular social groups.

18. Crehan 1997, 30. By "normal" ways of acting, we refer to the social construction of a shared sense of what behaviors are "appropriate" and right in a given situation. Our ideas draw from recent work on culture in anthropology, sociology, geography, and history.

19. This does not mean that everyone who shares a lifeworld has an identical and congruent worldview and set of values or that people do not sometimes question

their own (and one another's) premises. Rather, it means that they share important beliefs and assumptions about how the world works, although they may also hold other beliefs and assumptions that are different or even contradictory. At the same time, some of this variation may be quieted by the power of group norms. That is, lifeworlds also have a certain disciplinary element to them, in that some ideas within a particular lifeworld are not simply "unthinkable" but are effectively disallowed. Thus, our conceptualization differs somewhat from Schutz's and Habermas's in the sense that the content of a particular lifeworld is not simply or only in the "background" but is also explicit.

20. Government regulators are a fourth group of actors who were also important to the biotechnology struggle and who also figure into our story, albeit to a lesser extent than activists, corporate executives, and industry scientists.

21. Quoted in Charles 2001, 95.

22. Telephone interview with John W. Hanley, February 17, 2006.

23. The argument we are making here about the culture of science is consistent with the science and technology studies (STS) perspective, which views science not only as a set of practices but also as deeply cultural and involving particular belief systems.

24. See McAdam, McCarthy, and Zald 1996; Kitschelt 1986; Rucht 1990; Brocket 1991; Kriesi et al. 1992; and Tarrow 1989.

25. On the importance of culture, see Gamson and Meyer 1996; Klandermans and Johnston 1995; Nelkin and Pollack 1981; and Jenson 1995.

26. The term *global commodity chain* refers to the actors and relationships involved in producing, transforming, distributing, and consuming a particular commodity. The literature on global commodity chains and related concepts (e.g., global value chains, global supply chains, global production networks) is now vast. For an excellent overview, see Bair 2008.

27. Industries that produce products having multiple uses, such as plastics and rubber, are not as vulnerable to the actors involved in any particular chain, because they can easily switch to different relationships and product streams.

28. For an important exception, see Rachel Einwohner's work on the culture of animal rights activists and their adversaries (Einwohner and Spencer 2005; Einwohner 1999).

29. Erica Schoenberger (1997) makes this argument compellingly. For similar arguments about the cultural world of economic agents, see Ho 2009 and Moreton 2009. For recent collections that expound upon the inextricability of the cultural and economic spheres, see Fisher and Downey 2006; Callon 1998; Amin and Thrift 2004; and Du Gay and Pryke 2002.

30. As noted earlier, culture is a central dimension of our lifeworld concept.

31. Schoenberger 1997, 122.

32. In the United States, where the general public is accustomed to a high level of firm aggressiveness, and industry prowess is often the object of admiration as a

sign of strength, the industry's extreme aggressiveness paid off. This same strategy would not have worked in Europe, however, which is perhaps in part why the biotechnology and food industries did not engage in similar behavior there. Other reasons existed as well, related to the contingent political context of food scares.

33. Tarrow 2005.

34. Rootes 2003; Tarrow 2005.

35. We adopt the term *relational comparison* from Gillian Hart (2002), who uses it in a somewhat different context.

36. The information in this section comes from fieldnotes taken while participating in these two conferences, as well as from the materials collected at them.

37. The official title was the "7th International Gathering on Biodevastation: Genetic Engineering: A Technology of Corporate Control: A Forum on Environmental Racism, World Agriculture, and Biowarfare." The Gateway Green Alliance in St. Louis, the Institute for Social Ecology in Vermont, and a number of other small, left-leaning U.S. foundations sponsored the conference.

38. It was clear that the organizers of the World Agricultural Forum sought to include some NGO voices in its list of official invitees. Nonetheless, the scores of agribusiness representatives, heads of government and state agricultural agencies, officials from international development banks and large aid agencies, and others so outnumbered the handful of NGO representatives there that the last group was virtually invisible.

39. Author interview, August 7, 2007.

1. Precursors to Protest

1. Though scholars have theorized these developments in a variety of ways, they generally depict a postindustrial economic system in which productive relations and indeed the logics of accumulation are qualitatively distinct from those that exist under industrial capitalism. In this system, leadership is based on the control and management of informational technologies (including information on the "genetic codes" of living organisms) rather than on industrial capacity, and the centrality of capital–labor relations in economic organization is displaced by the importance of knowledge and education.

2. Bud (1998) argues that a significant shift in prominence took place within biology from microbiology to molecular biology in the postwar science establishment. Genetics also gave way ultimately to what appeared to be the revolutionary (but quite speculative) potential of genomics.

3. Wade 2007.

4. Kay 1998.

5. Krimsky 2003; Markle and Robin 1985.

6. The quoted phrase is from Jasanoff 2005, 34.

7. Szasz 1994.

8. Eisner 2000.

9. Dawkins 1993.

10. The "principle of science" aimed to provide a procedural basis to determine the legitimacy of the WTO's Sanitary and Phytosanitary Measures (SPS) Agreement, which dealt with food safety. Article 5.1 of the agreement made it a fundamental obligation to base such measures on a scientific risk assessment. Part of the argument for this was that it would narrow information gaps between exporters and importers and stabilize expectations among trading partners.

11. From the outset, the TRIPS agreement was controversial. Since most of the "innovation expertise" the agreement promoted lay in the northern countries, many in the global South feared it would lead to a massive transfer of capital from southern countries to the global North. Developing countries also noted that areas in which they had a competitive advantage, such as commercially usable traditional knowledge and folklore, received weak protection under TRIPS. For a comprehensive review of TRIPS, see Bellmann, Dutfield, and Meléndez-Ortiz 2003.

12. McMichael (2000, 254) defines the Globalization Project as "an emerging vision of the world and its resources as a globally organized and managed free trade/free enterprise economy pursued by a largely unaccountable political and economic elite." For a brief breakdown of its components, see McMichael 2000, 187.

13. Business and Industry Advisory Committee to the OECD 1998.

14. Jasanoff 2005.

15. Beck 1992.

16. Author interview, February 21, 2002.

17. Access of civil society organizations to the UN's multilateral deliberations on the environment had started as early as the first UN-sponsored conference, on the human environment in Stockholm in 1972. Such access was facilitated by the United Nations Environment Program, which was established after that conference. In 1996, the UN made formal "consultative status" available to international NGOs.

18. Buttel 2003.

19. Prudham 2003, 80.

2. Creating an Industry Actor

1. An *organizational field* refers to "those organizations that, in the aggregate, constitute a recognized area of institutional life: key suppliers, resource and product consumers, regulatory agencies, and other organizations that produce similar services or products" (DiMaggio and Powell 1983, 148).

2. These two phenomena are of course related; see Schoenberger 1997 for more discussion.

3. This discussion relies heavily on Martin Kenney's excellent study of the

biotechnology industry's early years (Kenney 1986); it also draws on Kloppenburg 2004.

4. Kenney 1986, 140.

5. Fikes 1999; Martineau 2001.

6. Fikes 1999.

7. See Martineau 2001 and Charles 2001 for a rich sense of the challenges involved and the intensity of the competition.

8. Author interview, February 8, 2006.

9. Author interview, February 17, 2006.

10. Author interview, February 21, 2006.

11. Quoted in Dickinson 1988, 2.

12. DuPont 1982. This thirty-three billion dollars refers to total sales.

13. Abbott Laboratories 1983.

14. This discussion is based on Kenney 1986.

15. National Academy of Sciences 1994.

16. *Chemical Week* 1983, quoted in Kenney 1986, 213.

17. Dickinson 1988.

18. Author interview with former Monsanto official, March 28, 2006.

19. Such contracting arrangements also benefited biotechnology start-ups by providing them with much-needed capital to help fund their research.

20. Actually two groups collectively forged this new career path for the biological sciences. One comprised scientists who were already professors and had achieved some prominence in their fields; the people in this mid- to late-career group were the ones who decided to start their own biotech companies or were recruited into industry to head up corporate research programs. The other group comprised the new generation of freshly minted scientists, whose typical career path in the past would have been a university postdoc and then a job as an assistant professor (Kenney 1986). For many scientists, old and young, breaking the mold was a difficult decision (see Martineau 2001).

21. In the late 1980s, Ralph Quatrano returned to academia, where he became chair of the biology department at Washington University, in St. Louis, and holds a distinguished professorship.

22. Author interview, June 8, 2004.

23. Mary Dell Chilton, quoted in Kloppenburg 2004, 191.

24. Author interview, July 10, 2004.

25. Author interview, July 9, 2004.

26. Author interview, July 10, 2004.

27. Author interview, February 21, 2006.

28. Author interview, June 8, 2004.

29. Author interview, February 8, 2006.

30. Author interview, July 9, 2004.

31. Quoted in Rotman 1993, 33, and cited in Boyd 2003.

32. Author interview, February 8, 2006.

33. Author interview, March 3, 2006.

34. Even though these two perspectives came together and became mutually reinforcing, this process was not seamless. In most large conglomerates, some of the "old guard" had to be convinced of the value of their company's shift toward molecular biology. This was especially true of agricultural chemical companies, which had been researching, developing, and marketing agrochemicals for years. When agricultural biotechnology came along, some employees viewed it as a threat to the products in which they had invested so much time and effort. (See Charles 2001.)

35. The views discussed in this section generally apply both to the professional scientific and business staffs of these companies alike.

36. Author interview, July 10, 2004.

37. Author interview, June 8, 2004.

38. The one important exception to this rule involved the science-based concerns raised by other concerned and outspoken scientists, such as Professor George Wald, of Harvard. While these arguments were not summarily dismissed, they did not gain any real traction when they were made in the mid-1970s. One possible reason was that the Watson–Crick dogma had yet to be debunked and remained the dominant way of understanding gene transfer; a second was that the tide of biotechnology was already so strong that the few scientists who objected were effectively whistling in the wind. We thank Dr. Robert Archer for calling our attention to this point.

39. Author interview, February 26, 2006.

40. Author interview, July 10, 2004.

41. Author interview, July 9, 2004.

42. Author interview, July 9, 2004.

43. Author interview, March 3, 2006.

44. Ho 2009; Reich 2007.

45. Author interview, June 8, 2004.

46. This information comes from the Ciba Specialty Chemicals Web site, at http://www.cibasc.com/index/cmp-index/cmp-about/cmp-abo-history.htm.

47. In April 2006, Novartis acquired Chiron outright, after an eleven-year partnership.

48. See Novartis 2006. In 2000, the company split its agricultural businesses off from its pharmaceutical and medical businesses. The agricultural component of the operation was renamed Syngenta.

49. Schmidta and Rühli 2002; Novartis 1997. Sandoz was originally established as a dyestuffs factory in Basel in 1886. See the company's Web site for its history. The number of countries in which Sandoz operated was calculated from the company's Web site on corporate affiliations.

50. Many scientists were first exposed to the practice of patenting in graduate school, at least if they were trained after the 1980 passage of the Bayh-Dole Act.

For the first time in U.S. history, the Bayh-Dole Act allowed universities to exclusively license the inventions they had made with federal funds to other parties, thereby securing a stream of income from publicly funded university research. For more information on this act, see "The Bayh-Dole Act" on the Web site for the Consumer Project on Technology, at http://www.cptech.org/ip/health/bd/.

51. Author interview, April 6, 2006.

52. Author interview, March 3, 2006.

53. Author interview, February 21, 2006.

54. How much a large biotech company thinks is "enough" is useful for benchmarking. Later in our conversation, this scientist noted, "Let's say it costs ten million dollars in research to generate this product. That's actually peanuts. Because we're looking at making a hundred million to two hundred million dollars a year on the product. If that business guy doesn't come back and say—if he only says, 'You're going to make thirty-five million dollars or fifty million dollars,' it ain't worth it" (author interview, February 21, 2006).

55. Oligopolies typically emerge when the barriers to entry or the cost of being in a business or both are very high, which is precisely the case for these science-based industries. One of the entry barriers in the agricultural and medical biotechnology industries is the relatively high costs of R&D, as already suggested. Another, even more formidable cost is that of moving a product through the testing and regulatory processes and then marketing it.

56. We say "most" because not all firms adopt this strategy of competition; some seek to compete on the basis of finding loopholes in intellectual property rights (or simply violating them entirely) and root their economic success in a "copycat" strategy.

57. As of this writing, Monsanto remains the leader in the agricultural biotechnology industry, ranking first in market share of genetically modified seed sales. DuPont ranks second, and Syngenta, third.

58. Schoenberger 1997, 150.

59. Levinson and Rosenthal 1984.

60. Author interview, March 3, 2006.

61. Author interview, March 3, 2006.

62. Author interview, June 7, 2004.

63. This discussion is based on interviews with two former Monsanto officials and on the investigative journalism of Eichenwald, Kolata, and Peterson 2001.

64. Author interview, March 3, 2006.

65. This observation is based upon our own interviews with former company employees and on other accounts of Monsanto culture, including Charles 2001.

66. Author interview, February 21, 2006.

67. Story told in Charles 2001, 38. Charles offers another amusing anecdote about Mahoney's growing impatience with the slowness of the R&D process in the 1980s. Pulling the company scientists into a meeting, Mahoney told them, "You

know, I've just come back from Germany, and I've got a metaphor for technology research. We were driving down the Autobahn at about 150 miles an hour toward Frankfurt. Every 20 miles or so there was a sign for an exit ramp that said *'Ausfahrt'* [German for 'exit']. That reminds me of our R&D. We're barreling along with all this expensive equipment, but we have to have a product now and then! An *Ausfahrt!*" (Mahoney, quoted in Charles 2001, 38).

68. Roger Malkin, quoted in Charles 2001, 123.

69. Author interview, June 7, 2004.

70. Various author interviews; Charles 2001; Eichenwald, Kolata, and Peterson 2001.

71. This was one of the reasons the company supported federal regulations in the United States and worked hard to derail states' efforts to establish state-level regulation (author interviews with former Monsanto officials, 2004). State-level regulation took more resources to manage and was politically harder to control.

72. An executive from one of Monsanto's competitors described Monsanto as having five times as many regulatory personnel as his firm did; even if he was exaggerating, the point is clear.

73. Author interview, May 22, 2004.

74. Author interview, February 21, 2006.

75. Author interview, July 10, 2004.

76. Magretta 1997; Shapiro 1998.

77. Author interview, July 8, 2004.

78. Charles 2001; World Resources Institute 2001.

79. Monsanto Company 1999.

80. Author interview, February 19, 2002.

81. Tom Urban, quoted in Charles 2001, 158.

3. Forging a Global Movement

1. Fowler et al. 1988.

2. Bogève Declaration 1987.

3. Social movements can adopt a variety of organizational forms, ranging from mass movements, in which many thousands of people are mobilized, to movements that are more networklike in nature and take the form of many loosely connected groups working on the same issue. The anti-biotech movement took the latter form.

4. Rochon 1998, 8.

5. Ibid., 15.

6. With *grievance formation,* we refer to the process of identifying some condition in society as a social problem and something that needs to be changed; an obvious example is racism. Grievances form the basis for social movements, in that they provide a motivation for organizing.

7. Ibid., 22.

8. Ibid., 25.

9. Eyerman and Jamison 1991.

10. With *capitalist commodification process,* we refer to the process by which goods, services, human labor, and so forth, which have not historically been exchanged in the market, begin to take on a market value and to be bought and sold.

11. Jasper 1997, 12.

12. Author interviews with anti-biotech activists, February 21, 2002, and June 24, 2002.

13. Author interview, June 13, 2002.

14. Perlas 2003. In the early 1980s, Perlas worked with Jeremy Rifkin on a book titled *Algeny.* After returning to the Philippines, he continued working sporadically on the issue of agricultural biotechnology.

15. This group included the Nobel laureate James Watson, three future Nobelists (Paul Berg, David Baltimore, and Daniel Nathans), and the bacterial geneticist who shared the first patent on rDNA, Stanley Cohen (Newman 2001).

16. Krimsky 1982.

17. Moore 2008; Zimmerman et al. 1972.

18. National Academy of Sciences 1977.

19. Cavalieri 1976.

20. Author interview, October 26, 2004.

21. National Academy of Sciences 1977, 40.

22. Author interviews with Francine Simring, October 2, 2006; Sheldon Krimsky, January 23, 2002; and personal communication with Stuart Newman, August 2006.

23. Committee for Responsible Genetics 1983.

24. The first issue of *GeneWatch* contained articles on the EPA and regulations, genetic screening in the workplace, biological weapons, and "oncogenes" and lab safety, as well as seeds and biotechnology (ibid.). Subsequent issues included articles on biotechnology and the third world, prenatal screening, and the "deliberate release" of GMOs and their environmental risks, among other topics.

25. Author interview, June 20, 2002.

26. Howard and Rifkin 1977.

27. Author interview, June 20, 2002.

28. A decade later, Haerlin would become head of Greenpeace International's campaign against genetically modified food and a key strategist of its Europe campaign. In October 2006, the Gen-ethisches Netzwerk celebrated its twentieth anniversary. (See http://www. gen-ethisches-netzwerk.de/.)

29. Lappé, Collins, and Fowler 1977.

30. See Fowler 1978. Although it was a small booklet published on a shoestring budget, *Fowler's Graham Center Seed Directory* made quite a splash when it came out.

31. Author interview, June 13, 2002.

32. Author interview, August 15, 2006.

33. Fowler, cited in the Peoples Business Commission Amicus Brief 1979.

34. See Kloppenburg 2004 and Boyd 2003 for detailed analyses of historical changes in U.S. intellectual property rights law and policy.

35. Mooney 1979.

36. GRAIN 1993.

37. Author telephone interview with Pat Mooney, August 15, 2006.

38. Perlas originally presented his analysis at a Penang conference organized by the Asia-Pacific Peoples Environmental Network and Sahabat Alam Malaysia (Friends of the Earth–Malaysia) in 1987 (Perlas 1994).

39. Oliver and Johnston 2000.

40. Eyerman and Jamison (1991), Rochon (1998), and Tesh (2000) have all made related arguments in the context of other social movements.

41. Author telephone interviews with Ruth McNally, November 7, 2004; and Peter Wheale, January 20, 2005.

42. Author interview, February 15, 2004.

43. McAfee 2003.

44. Eyerman and Jamison 1991, 57.

45. Author interview, June 19, 2002.

46. Author interview, November 18, 2004. At the time, both Kloppenburg and Kenney were doctoral students at Cornell University. Kloppenburg was working on a history of plant breeding and the "commodification of the seed," and Kenney was studying the commercial development of the biotechnology industry and its impact on university–industry relations. Kenney's book, *Biotechnology: The University–Industrial Complex,* came out in 1986, and Kloppenburg's book, *First the Seed: The Political Economy of Plant Biotechnology,* first appeared two years later. As students, Kenney and Kloppenburg had also published several other pieces that were in circulation as early as 1984. Both had attended a number of activist meetings, where they shared their ideas and learned from those present.

47. Author interview, October 26, 2004.

48. Purdue 2000.

49. Author interview, July 7, 2000.

50. Interviews with Jeremy Rifkin, 2001; and Benny Haerlin, December 2004.

51. Snow and Benford 1992, 137.

52. Author interview, June 20, 2002.

53. Fowler et al. 1988, 29.

54. The Cartagena Protocol on Biosafety went into force on September 11, 2003. In chapter 6, we discuss activists' role in getting the protocol negotiated.

55. We are not the first to make the claim that activism is motivated by moral concerns. See, for instance, Jasper 1997; Luker 1984; and Smith 1996.

56. This section is based on the authors' personal interviews with activists

from the anti-genetic-engineering movement, which will not be individually cited here. These interviews were conducted from 2001 to 2004.

57. Author interview, June 20, 2002.

58. Levidow 1999; Purdue 2000.

59. Purdue 2000.

60. Ganz 2000.

61. Author interview, June 20, 2002.

4. The Struggle over Biotechnology in Western Europe

1. This industry worldview also made it difficult for the whole of the U.S. agricultural biotechnology industry to grasp the significance of different state–society relations, particularly at a moment when those relationships were changing as European countries struggled to work out their identities in the context of an emerging European Union (Jasanoff 2005; Tiberghien and Starrs 2004).

2. This section draws on Gottweis's (1998a) and Jasanoff's (2005) excellent comparative studies of biotechnology policy in the United States, Germany, and Britain.

3. Gottweis 1998a, chapter 4.

4. Ibid.

5. On French policy toward biotechnology and the ensuing controversy, see Marris 2000; and Bonneuil, Joly, and Marris 2008.

6. Kahn 1996, cited in Marris 2000.

7. The law is known as the 1986 Act on Gene Technology and the Environment; in its original form, it contained strict limitations on the deliberate release of GMOs and on the introduction of GM food, which required government approval. See Jelsoe et al. 1998.

8. For reasons of space, our narrative focuses on activism in a very limited number of countries in this early period. Our decision to focus on Germany and Britain derived from our greater familiarity with these cases and our greater access to primary sources. Activism was also strong, however, in France, Austria, and several other countries.

9. This discussion draws heavily on Gottweis 1998b and Jasanoff 2005. We also utilize information collected in personal interviews with German activists.

10. Jasanoff 2005.

11. The resources and work that went into the Enquiry Commission were remarkable. Its three-year project included the establishment of seven working groups to address different potential fields of application and the solicitation of scores of expert opinions. The commission was composed of eighteen members (only one of whom was from the Green Party) and had nine full-time employees, including five scientists.

12. Ibid.

13. Enquiry Commission on the Prospects and Risks of Genetic Engineering, 1987, quoted in Jasanoff 2005, 60.

14. Jasanoff 2005.

15. Ibid.

16. Both Gottweis's and Jasanoff's work suggests that during the 1970s, the dominant policy narrative about biotechnology was that assessment of its risks were best left up to scientific experts, who were also capable of controlling those risks, that is, regulating themselves. Gottweis (1998a, chapter 6) argues that a new policy narrative took form in the 1980s, when activist dissent, unfavorable media coverage, and an increasingly skeptical public created a serious "crisis of governability" around biotechnology and a high level of uncertainty for the industry. This growing sense of crisis prompted the state to develop a new regulatory regime for biotechnology and a new dominant narrative, and these led scientists and the industry to support it.

17. This discussion draws on historical–archival material and telephone interviews with Joyce D'Silva, staff member of Compassion in World Farming, a British NGO; Dr. Tim Lang, professor of food policy at City University, London; and Dr. Eric Brunner, a biochemist and social epidemiologist based at University College London. We thank Susan Pastor for providing us with access to her personal files, which included early correspondence between the anti-bST campaigns in Britain and Wisconsin. Purdue 2000 offers a more comprehensive discussion of early British activism.

18. According to Lang, the London Food Commission's concerns about bST were threefold: human health impacts, animal welfare impacts, and the likelihood that bST would promote economic concentration in the dairy sector, benefiting some farmers through higher output while forcing others (particularly smaller dairies) out of business.

19. Brunner 1988.

20. See Millstone, Brunner, and White 1994.

21. The European Commission, the executive arm of the EU, is the only one of the three EU institutions (the commission, the European Council, and the European Parliament) that has the power to initiate legislation. Originally, the European Parliament was the weakest of the EU institutions, though its power increased substantially after the Maastricht Treaty was signed in 1993.

22. Crespi 1999.

23. These included the European Patent Convention, established in 1973, and each country's own national patent system (see Leskien 1998).

24. The information in this paragraph comes from Emmott 2001 and an author interview with Emmott. Emmott was the policy adviser on genetic engineering for the Green Group of members of the European Parliament in the late 1980s and 1990s and a long-time participant in the campaign against the directive.

25. A considerable dispute occurred among activists, the biotechnology industry,

and policymakers over the meaning of this directive and the significance of the changes it signified from existing national and European Patent Office law and policy. From the activists' viewpoint, the directive profoundly augmented the range of products, processes, and information that could be patented. From the industry's and many policymakers' perspectives, the directive merely formalized what was *already* patentable under existing patent laws and conventions and made patent protections in EU countries consistent. While the true significance of the directive will become clear only over time, it must have offered some perceived benefits to industry, or industry would not have actively supported it.

26. According to one source, the Parliament presented some forty proposed amendments to the text, many of which were activist-inspired and introduced by Green Group parliamentarians. See GRAIN 1995.

27. Ibid.

28. See Jasanoff 2005, chapter 3. Patterson (2000) makes a similar argument but goes so far as to suggest that four Directorates-General had strong interests in biotechnology policymaking. In addition to those already mentioned, she identifies DG III (Enterprise and Industry) and DG VI (Agriculture and Rural Development) as having legitimate claims for being intimately involved. Ultimately, DG III and DG XII did play a significant but not a controlling role.

29. Author interviews with European activists from the Greens and Friends of the Earth–Europe, 2004 and 2005.

30. See Patterson 2000 for a detailed discussion. In the United States, a product-based system based on the assumption of substantial equivalence won the day in the 1980s. Several analysts of GMO politics have argued that this difference in regulatory approaches helps to explain why GMOs have had such different fates in the United States and Europe (Bernauer 2003; Jasanoff 2005).

31. In 2001, Directive 90/220 was replaced by an even more restrictive set of rules (Directive 2001/18). Among the specific changes made to the framework were the addition of a new requirement that all GM products be labeled before they could be approved and a provision that member states would have to assume responsibility for ensuring the "traceability" of GMOs at all stages. See Carr 2000.

32. Author interviews, November 2004, January 2005.

33. Jasanoff argues that the Environment Directorate rejected the charge of "Green capture," suggesting that DG XI's policy designers simply shared some of the activists' concerns about biotechnology, and favored an approach characterized by safeguards and precaution rather than assumptions about the technology's benign character (2005, 82). While this was quite likely true, the two phenomena are not mutually exclusive; indeed, given the opposing pressures coming from some arenas of government and from the biotechnology industry, such a process-oriented stance on the part of the Directorate was likely strengthened by the Greens' and other NGOs' positions.

34. The *Bt* gene enables a plant to produce its own toxin, thus creating an

internal form of insect resistance; in official U.S. regulatory terminology, *Bt* is referred to as a "plant incorporated protectant."

35. This story was told to us by a colleague of Haerlin's.

36. Bonneuil, Joly, and Marris 2008.

37. Kempf 2003.

38. These figures are from http://www.cjd.ed.ac.uk/figures.htm, accessed April 28, 2003.

39. Thomas 2001.

40. Ibid.

41. Bové and Luneau 2002, cited in Bonneuil, Joly, and Marris 2008.

42. Schweiger 2001.

43. Cited in Marris 2000.

44. Jelsoe et al. 2001.

45. Levidow 2000.

46. Eurobarometer survey data for 1996 indicate that when Europeans were asked to select the institutions best positioned to regulate biotechnology, the most frequent answer was "international institutions such as the UN and the World Health Organizations" (34 percent of respondents). Respondents ranked "scientific committees" second (22 percent) and "national public bodies" third (12 percent). In a subsequent question, respondents were asked in whom they had confidence "to tell them the truth about GM crops grown in fields." Only 4 percent answered "national public bodies," and only 1 percent answered "industry." A comparable survey carried out in the United States in 1997 showed very high levels of public trust in the FDA and USDA: 84 percent of survey respondents claimed they trusted the FDA "some" or "a lot," and 89 percent felt the same way about the USDA (Gaskell, Thompson, and Allum 2002).

47. This story is told in Charles 2001.

48. Schweiger 2001.

49. Activists lodged a long list of complaints to the British Advertising Standards Authority about the veracity of the claims Monsanto was making in its advertising campaign. The ASA upheld four of the thirteen complaints made about Monsanto's full-page advertisements, and the issue was widely reported in the press (BBC News 1999b).

50. Interview with Juniper, March 16, 1999, on http://www.biotech-info.net/npr_monsanto.html, accessed July 12, 2007.

51. According to one former Monsanto employee, Monsanto relied quite heavily on this type of pressure during Shapiro's leadership (interview with former Monsanto employee, June 2004.)

52. Shapiro, quoted in Specter 2000, 58.

53. Gaskell, Allum, Bauer, et al. 2000, 938.

54. Ibid., Table 2.

55. Data are presented in Gaskill and Bauer 2001, cited in Bernauer 2003, 92.

56. Greenberg 1998.

57. *The Economist* 1998, 79–80.

58. Quoted in Charles 2001, 222.

59. Gaskell, Allum, Bauer, et al. 2000.

60. Henson and Northern 1998.

61. This discovery was announced March 1996, "trigger[ing] a crisis of confidence in the manner beef was produced and regulated among both citizens and regulators" (Vos 2000, 232).

62. Quoted in Lagnado 1999.

63. While the particular details of the risk assessment are not important for our story, the point is that the scientific data required by the risk assessment left plenty of room for a country to object that the information provided was insufficient and to ask for more, thereby holding up approval, or even consideration, of the dossier.

64. European Council 1990.

65. Directive 90/220 refers to the Council Directive of April 23, 1990, on the deliberate release of genetically modified organisms into the environment.

66. See European Council 1990, available on the Web at http://www.biosafety. be/GB/Dir.Eur.GB/Del.Rel./90.220/annex2.html#Pt%202.

67. According to one activist–informant, more than a dozen petitions were approved before the activist community "got up to speed on it" and started working to slow the approval process down (author interview with European activist, November 22, 2004).

68. Telephone interview with activist involved in the process, November 22, 2004. According to this individual, careful scientists had no difficulty in finding weaknesses in the risk data that most firms provided in their dossiers, because firms were reticent to disclose much information, especially information they saw as proprietary.

69. Ibid.

70. This pressure had different roots and manifestations in the various European countries. In Britain, a key source of this pressure came from the government's disastrous handling of mad cow disease. In France, the pressure came from a growing number of public protests against GMOs, the criticism that some prominent ecologists and population geneticists had begun to level at the government for relying too heavily on molecular biologists for its risk and safety assessments rather than considering the opinions of other branches of science, and several controversial court cases involving GMOs (Joly and Marris 2003; Marris et al. 2005). In Germany, it came from the political mobilization of the 1970s.

71. Author interview, November 22, 2004.

72. Marris 2000.

73. Technically, the commission must make a decision on the validity of the state's justification within a three-month period, and the member state is obliged

to abide by that decision. In fact, the commission has not held maverick states to this rule. See ibid., 33.

74. That GM corn was Novartis's *Bt* 176; see Bernauer 2003, chapter 5; and Patterson 2000, 338. On the French case, see Marris 2000.

75. Jasanoff 2005.

76. Marris 2000.

5. Creating Controversy in the United States

1. The statistics on GM crop use are from USDA/ERS 2002.

2. Wright 1994.

3. In 1976, the National Institutes of Health adopted the recommendations that came out of the Asilomar conference and created a four-tiered system of classifying rDNA research based upon its perceived biohazards.

4. This discussion draws on Culliton 1976; and Goodell 1979.

5. Goodell 1979.

6. Quoted ibid., 38.

7. Krimsky 1991.

8. Palca 1986.

9. The following account draws on Elmer-Dewitt 1987; Krimsky 1991; and U.S. Congress 1988.

10. This discussion relies on the more detailed accounts in Krimsky 1991; and U.S. Congress 1988.

11. The story is more complicated than we have painted it here; see the studies cited above for full accounts.

12. Schacter 1999, 42. Actually, not all research on this ice-nucleating microbe was stopped. The company that bought AGS ended up developing and marketing a naturally occurring *P. syringae,* which they then licensed to Ecogen, a small biopesticide company.

13. According to Adam Sheingate, a January 1984 report carried out by the Congressional Office of Technology Assessment called for decreased regulation in order to maintain U.S. dominance in biotechnology. Sheingate notes that the White House established a cabinet council working group on biotechnology to coordinate regulatory authority. On the basis of interviews with House Agricultural Committee staff, Sheingate argues that the White House officials leading this working group sought to "avoid new regulations and limit EPA authority over the testing and marketing of commercial biotechnology products [and] . . . encouraged the FDA and USDA to assert their role in biotechnology based on their existing statutory authority over food and crops" (Sheingate 2006, 249).

14. The information in this paragraph comes from Eichenwald, Kolata, and Peterson 2001 and from the authors' interviews with two former Monsanto officials. The first two products that landed on the Reagan administration's doorstep were

a recombinant form of bovine growth hormone, produced by Monsanto, and the Flavr Savr tomato, developed by the California company Calgene. Both firms sought regulatory approval in the 1980s and received it in 1993 and 1994, respectively.

15. Our interviews with two former Monsanto officials involved in the regulatory push suggest that different companies had different opinions on the regulatory question, with some firms strongly desiring regulation and other firms deeply opposed. As the most powerful player in the industry, Monsanto felt it did not need to strike a compromise position with those opposed to regulation but went directly to the top and pushed the EPA and FDA for regulations (author interviews with Monsanto officials, June 2004; see also Eichenwald, Kolata, and Peterson 2001).

16. *Federal Register* 49, 50856 (December 31, 1984), quoted in Kelso 2003, 243.

17. Quoted in Kelso 2003, 243.

18. The Coordinated Framework was first published in proposal form in the *Federal Register* in December 31, 1984, and revealed clear differences in approach among the three agencies. The EPA favored a process-based approach to regulating products of biotechnology, whereas the other two agencies, especially the FDA, preferred a product-based approach. The final version established a product-based policy, suggesting that the FDA and the USDA (and the major biotechnology companies) had won out (Sheingate 2006).

19. *Federal Register* 57, 6753 (February 27, 1992). In other words, regulations would be imposed only if a real and substantial risk, not a small or hypothetical risk, could be proved. See Kelso 2003.

20. If a food was known to have certain characteristics, such as being an allergen, it was not considered substantially equivalent, or "GRAS" (generally recognized as safe). In this case, it would need to go through an FDA review. This discussion is based on Sheingate 2006, 253–54.

21. For a revealing look into how key FDA administrators viewed regulation, see the writings of biotech regulator Henry Miller (1997a, 1997b).

22. These data come from Shoemaker 2001.

23. Estimates of how much these four chemical and drug giants spent range from one hundred million dollars to eight hundred million; our best guess is that the money invested directly in this project was on the lower end of this range (Ingersoll 1990).

24. Schneider 1990b.

25. Interestingly, this reading of the data was not unanimous within the FDA. An administration veterinarian, Richard Burroughs, took issue with his superiors' interpretation of the animal health evidence, causing quite a stir within in the administration. In 1988, the administration relieved Burroughs from his rBGH-related duties and soon thereafter fired him. According to journalist Bill Lambrecht, Burroughs alleges he was let go because he wanted more data from the companies and was not willing to back down when they fought back (Lambrecht 1990a).

26. Tangley 1986.

27. Jeremy Rifkin, quoted in Schneider 1988.

28. Tangley 1986.

29. Cannell, quoted in Schneider 1988.

30. Quoted in McNair 1990. Because rBGH is considered an animal drug, it is regulated by the FDA.

31. On the basis of information obtained under the U.S. Freedom of Information Act, FoET's suit charged that the National Dairy Promotion and Research Board had worked with the four rBGH manufacturers to promote the hormone's acceptance by the public, even though rBGH had not been proven safe and its adoption was not in the interest of all U.S. dairy farmers. See McCormick 1989.

32. *Los Angeles Times* 1989.

33. Thornley 2001. At the time, Dr. Samuel Epstein was professor of occupational and environmental medicine. For the last twenty years, Epstein has been the chief critic of the human health risks associated with rBGH milk.

34. Richards 1989.

35. In February 1994, the FDA issued an interim opinion on labeling of BGH milk, which permitted farmers and stores to label their milk as being made from cows that were not treated with rBGH (or as it is also called, rBST). However, any farmer or dairy that chose to label also had to include the disclaimer that "no significant difference has been shown between milk derived from rBST-treated and non-rBST-treated cows." In 1996, the courts ruled that the mandatory labeling law was illegal. See Lambrecht 1991.

36. Schneider 1990a.

37. Richards 1989.

38. Lambrecht 1989.

39. Quoted in Richards 1989. So zealous did Monsanto become that the FDA found cause to reprimand its behavior. In January 1991, the FDA sent Monsanto a letter "charging that the company had illegally promoted BST as safe and effective in brochures, at scientific conferences and at meetings with farmers and consumers" (Lambrecht 1990b).

40. Interview with Robert Milligan, professor emeritus of agricultural economics at Cornell University, August 3, 2007. Milligan, who grew up on a dairy farm himself, works as a consultant to the dairy industry.

41. Quoted in Lambrecht 1990b.

42. Ibid.

43. Seppa 1989, 1990.

44. See, for instance, Thornley 2001. The companies and their lobbying organizations also used the courts to challenge states that sought to establish their own labeling systems. Vermont is a case in point; see Lambrecht 2001.

45. Adoption rates are difficult to establish because of a lack of national data and because most adopters do not use it on their whole herd. A review article by

Butler (in *AgBioForum*) suggests that a sizable minority of cows (perhaps on the order of 20–30 percent) were being treated with rBGH at the end of the 1990s. In 1999, Monsanto was the only company that was still marketing rBGH, under the trade name of Prosilac. See Butler 1999.

46. Shoemaker 2001.

47. Charles 2001; and author interview with former Monsanto official, St. Louis, June 7, 2004.

48. Author interview with David Tramel, Minneapolis, July 22, 2007.

49. Interviews with two funders and meeting attendees, January 23, 2002, and April 3, 2002.

50. GEAN was a grassroots-oriented network that ultimately came to include about two hundred activists and activist organizations in the United States; GEFA comprised seven established food and environmental organizations that agreed to work together on certain centrally planned actions.

51. Losey, Rayor, and Carter 1999.

52. Author interview, July 10, 2000.

53. On the industry's counterattack, see Lambrecht 2001, 79. In 2001, the National Academy of Sciences published a series of follow-up studies to Losey and his colleagues' work (Zangerl, McKenna, Wraight, et al. 2001).

54. Lagnado 1999.

55. Gillis 2000.

56. Ultimately, Aventis paid over ten million dollars to farmers in Iowa alone. And in March 2002, a federal judge settled a class action lawsuit by consumers who complained of allergic reactions to Starlink corn with an award of nine million dollars. This information comes mainly from Kaufman 2002.

57. The results of this survey are interesting. Of the thirteen choices they were given, the highest proportion of respondents named the FDA as the informational source that they trusted "a great deal" regarding GMOs. The second most trusted source was "friends and family" (37 percent trusted them "a great deal"), followed by farmers (34 percent), and the Environmental Protection Agency and "scientists and academics" (33 percent each). Environmental groups and consumer groups were trusted by 23 percent and 21 percent of respondents, respectively. These data are from http://www.americansworld.org/digest/global_issues/biotechnology (accessed September 15, 2007).

58. Zawel 2000, emphasis added.

59. Quoted in Barboza 1999.

60. These companies included Monsanto, Dow AgroSciences, DuPont, Syngenta, BASF, Bayer, and Aventis.

61. Gardner 2001.

62. Ibid.

63. In 2002, BIO had a budget of about thirty million dollars a year, employed

seventy people, and represented a thousand biotechnology companies working in the pharmaceutical, medical, agricultural, and industrial sectors (Elias 2002).

64. Cohen and Gillam 2002.

65. These figures are from the USDA's Economic Research Service's *Agricultural Outlook* report, at http://www.ers.usda.gov/publications/agoutlook/aug1999/ao263a.pdf (accessed August 1999).

66. The following discussion is based on the newspaper coverage of the issue, an interview with an organizer of the anti-GM wheat campaign (conducted June 18, 2007), and a chapter by Dennis Olson, a former organizer with the Western Organization of Resource Councils in Montana (Olson 2005).

67. Quoted in Kram 2001.

68. Rafferty 2001.

69. See ibid.

70. See Olson 2005, 152. According to Olson, the company and its allies launched a major lobbying effort to kill the bill in the Senate. See also Lyderson 2002. Wanzek is one of the featured speakers on the "benefits of biotech" on Monsanto's Web site, Conversations about Plant Biotechnology.

71. Theresa Podell, workshop presentation at the conference "Growing a GE-Free Northeast: Education, Strategy and Action against Genetic Engineering," November 1–3, 2002, Hampshire College, Amherst, Mass.

72. Like its counterparts in the other Great Plains states, the Dakota Resource Council was originally formed in the 1970s by farmers and residents of these rural communities to respond to the destructive effects of strip mining.

73. U.S. Wheat Associated Press Release 2002.

74. The Wisner report estimated that the foreign market loss for hard red spring wheat would be on the order of 30 to 50 percent and would be even more for durum wheat (Wisner 2003).

75. Cited in Olson 2005.

76. The range of groups in this coalition is quite striking including, among others, the Canadian Wheat Board, the Saskatchewan Association of Rural Municipalities, the National Farmers' Union, and Greenpeace. For an analysis of how such a diverse coalition of groups managed to work together, see Magnan 2007.

77. Ibid.; *Leader Post* 2002.

78. Pates 2002.

79. Lenz 2003.

80. Horner 2004.

81. It is worth noting that the first two of these pressures could, in part, be traced to activism.

82. Pharmacia did retain the agricultural chemical side of the business, however. Nonetheless, with Roundup about to go off patent, the company was about to lose the economic security associated with its leading agrochemical project.

83. Hindo and Schneyer 2007.

84. At the time of this writing (mid-July 2009), Monsanto's share price had dropped to seventy-five dollars a share.

85. By *first-generation,* we are referring to herbicide-tolerant crops, such as Roundup Ready corn and soy, and plants that are genetically engineered to have insecticidal properties, for example, *Bt* corn and cotton.

6. Biotech Battles and Agricultural Development in Africa

1. Paarlberg 2008, 23.

2. Zerbe (2004, 594) quotes Andrew Natsios, head of USAID, as proclaiming that anti-GM groups were "putting millions of lives at risk in a despicable way."

3. *Nature* 2008.

4. When South Africa approved *Bt* white maize in 2000, it became the first country in the world to approve a genetically engineered staple foodstuff.

5. For a definition and detailed discussion of transnational advocacy networks, see Keck and Sikkink 1998.

6. For discussions of these trends, see Beintema and Stads 2006; Eicher, Maredia, and Sithole-Niang 2006; and Keeley and Scoones 2003.

7. One interesting case is the Institute for Sustainable Development, an NGO set up in 1995 in Ethiopia by Tewolde Egziabher, who was the lead African negotiator in the Cartagena Protocol negotiations and became the Ethiopian minister of environment. The institute received support from the Gaia Foundation, in Britain, as well as the Third World Network in Malaysia.

8. In particular, as development analysts have noted, the World Bank and the International Monetary Fund have the effective power to determine a wide range of policy frameworks for poor countries, both through their aid conditionalities and their ability to provide expertise. More broadly, the World Trade Organization has the power both to make and to enforce rules among its members. As Wallach and Woodall (2004) note, only about 10 percent of the WTO document pertains directly to trade; the remainder sets constraints on domestic policy choices in domains as disparate as manufacturing and education. In either case, the sovereign capacity of states is substantially curtailed.

9. The notion of a "transnational opportunity space" was developed by Sidney Tarrow (2005) in his wide-ranging analysis of the "new transnational activism."

10. This process had a significant presence in Africa. For instance, the Nigerian NGO Environmental Rights Action was formed in 1993 as a local branch of Friends of the Earth–International to advance "environmental human rights." Although it focused mostly on the oil industry, the ERA/FoE became the principal civil society watchdog on debates about genetic engineering in Nigeria. In 1996, the African Biodiversity Network was established to link NGOs, civil society organizations, and community-based organizations across the continent. In line with

the broader international trend of boosting and "mainstreaming" civil society organizations, the network was funded by both the London-based Gaia Foundation and SIDA, the official Swedish aid agency. This network subsequently also developed national affiliations such as the Uganda Biodiversity Network and the Kenya Biodiversity Network, which became very active in mobilizing opposition to government efforts to introduce GMOs late in the first decade of the 2000s. For a discussion of other parts of the world, see Keck and Sikkink 1998, chapter 4.

11. Segarra and Fletcher 2001.

12. Abley 2000.

13. Author interviews with participating activists, June 2000, Minneapolis, August 2001; various press accounts.

14. Egziabher 2000.

15. Interview with an academic observer who requested anonymity, June 2000.

16. For a useful account, see Tjaronda 2006.

17. Gupta and Falkner 2006.

18. Buttel 2003.

19. See Egziabher 2003.

20. Yoon 2008; Clapp 2006.

21. Kuyek 2002; Weiss 2002.

22. In 1998, the USDA and Syngenta had taken out a patent on a gene technology that produced sterile crops, so that seed could not be saved and replanted. In the wake of an enormous outcry against this technology, the industry promised not to make it commercially available, though it reserved the right to change its mind. Activists, especially in India, had stoked the fires of fear regarding this technology, and many people still believe it is being disseminated. The term *Terminator technology* was coined by the activist group RAFI and proved to be a remarkably resilient and evocative frame. In a sense, the notion of Terminator technology did for the production end of GMOs what the notion of Frankenfoods did for the consumption end.

23. Kelemu et al. 2003.

24. Chinsembu and Kambikambi 2001.

25. APDB 2004, 7–8.

26. According to Keeley and Scoones (2003), the World Food Program seemed to have not even thought about the implications of importing GM food into countries that were participants in the Cartagena Protocol and had little or no regulatory capacity. The WFP certainly had no strategy for dealing with this complication.

27. For instance, Zimbabwe was an exporter of high-quality beef and was very reluctant to prejudice that market by inadvertently feeding its cattle on GM fodder.

28. Manda 2003, 5.

29. Mnyulwa and Mugwagwa 2003.

30. The World Bank (2007, 166–67) shows that between 1981 and 2000

agricultural R&D fell in sub-Saharan Africa from 0.84 to 0.72 percent of agricultural GDP. Although African countries did spend a greater percentage of agricultural GDP on R&D than other developing countries, they constitute the only area in the world where the percentage dropped over this period. See also Pingali and Raney 2005. For the dire situation in Zambia, see Elliott and Perrault 2006.

31. In this context, Egziabher (2003) argues that the distinction between "substantial equivalence" and the "precautionary principle" is one of principled philosophical differences in how to evaluate risk. Paarlberg, for his part, presents the distinction as one between proven risk and unproven or speculative risk.

32. Boudreaux 2006. The Monsanto South Africa Web site contains an article titled "Biotechnology: Biology of Hope." See http://www.monsanto.co.za.

33. Ofir 1994.

34. UPOV established a kind of patent right for plant breeders; see Kloppenburg 2004 for more detail.

35. Aerni 2002.

36. According to a senior scientist who was a founding member of SAGENE, the committee's members modeled their approach very closely on the approach laid out by the National Institutes of Health in the United States (author interview, July 30, 2007).

37. Woodward, Brink, and Berger 1999; South African Government 1997.

38. www.gmwatch.org/print-profile.asp?PrId=282 (accessed November 6, 2006).

39. South African Government 1997.

40. Indeed, in South Africa and other African countries, scientists have criticized the Cartagena Protocol for introducing unnecessary bureaucratic constraints on the effective flow of information, knowledge, and technology. Many have called for its abolition on these grounds.

41. See Gupta and Falkner 2006.

42. Mail & Guardian Online 2006.

43. Clark, Mugabe, and Smith 2002.

44. van Roozendaal 1995.

45. Quoted in Ashton et al. 2004.

46. This meant that a farmer who inadvertently grew GM crops would be liable to the company for infringing its patent, rather than the company being liable for allowing the farmer's fields to become contaminated. This provision considerably reduced the risk to biotech companies and may have played a part in encouraging companies to carry out illegal field trials in South Africa and Zimbabwe.

47. Cock 2007, 174.

48. See the Biowatch Web site at http://www.biowatch.org.za.

49. In 2007, SAFeAGE claimed to have some 250,000 members and 135 organizations in the network, as well as solidarity with a further 350,000 individuals internationally.

50. As expressed by an outreach facilitator for Biowatch who had helped organize a series of workshops for small-scale farmers: "I think in our government the officials are looking for a new, 'civilized' way of doing things and not actually looking at what kept society there for a long time without their intervention. What the government officials should be doing is to listen to the grassroots" (quoted in Hobbes 2009).

51. Author interviews, August 2007.

52. The quotation is from a pamphlet handed out by Safe Food Coalition activists at a protest outside the Health Department in Pretoria in August 1999. This was an especially alarmist event, at which activists distributed pamphlets linking genetically modified organisms to cancer, allergic reactions, and pollution of food and water. For an account, see Cook 1999.

53. Activists evinced two rationales for labeling. One was that consumers should be able to choose what they put into their bodies. The other was that when this technology went wrong (as many were convinced it would), liability could be assigned to its developers rather than to the "victims."

54. As we saw in chapter 4, conflicts over labeling had a profound effect on the politics of agricultural biotechnology in Europe. Similar conflicts have taken place across the globe and have helped to keep the technology controversial. In June 2008, the Conference of Parties to the Cartagena Protocol finally committed to set up a *binding* set of international rules on liability and redress of damage caused to health or environment by GMOs at its 2010 meeting. This decision will add teeth to the fight over labeling, since clear labeling will establish a route of traceability necessary for the apportionment of liability.

55. Biowatch appealed this decision all the way to the Constitutional Court of South Africa, on the argument that such decisions would have a chilling effect on poorly resourced civil society organizations acting in the public interest to press government for greater accountability and transparency. In June 2009, the Constitutional Court finally found unanimously in Biowatch's favor.

56. South African Government 1998, Section 2(4)(i).

57. Ashton et al. 2004, emphasis added.

58. In 2006, the regulatory authority accepted this argument and rejected the application. In 2008, however, it revisited the issue and approved greenhouse trials.

59. South African Government 2005, 5. In a 2007 national biotechnology audit, biotechnology companies declared their greatest constraint to be the time needed to gain regulatory approval (South African Government 2007, 23).

60. Scoones 2005, 10.

61. Quoted in Kameri-Mbote 2007.

62. The same thing has happened in India, where activists and farmer–activists have sought to mobilize farmers around the issue of what GM seeds will mean for them.

63. AGRA declares its first priority to be the development of better and more appropriate seeds. But its language makes clear that it will not privilege transgenic technology: "Alliance programmes are tackling these challenges through collaborative projects that bring farmers and scientists together to develop and distribute seeds suitable for local environments while also supporting genetic diversity and the rights of farmers to save seeds. The Alliance 'Programme for Africa's Seed Systems' (PASS) is funding African-led initiatives that are breeding new varieties of conventional maize, cassava, beans, rice, sorghum, and other crops to improve their resistance to disease and pests. The goal is to develop and release more than 1,000 improved crop varieties over the next ten years" (Alliance for a Green Revolution in Africa, n.d.).

64. As AGRA noted in launching its training program for crop breeders, "Most of the crops important to Africa—such as cassava, sorghum, millet, plantain, and cowpea—the so-called 'orphan crops,' are of little importance to researchers and educators in the developed world. As a result, there is a serious shortage of breeders of these crops. For example, there are under a dozen millet breeders in all of Africa. Yet millions of people in sub-Saharan Africa depend on millet as an important part of their diet. Conversely, most of the more than US$35 billion invested by private firms in agricultural research is concentrated in North America and Europe, on a handful of commercially important crops" (Alliance for a Green Revolution in Africa 2007).

65. Eicher and others have cautioned against "silver bullet" solutions, arguing that green revolutions take time, especially in the diverse and often fragile ecologies of Africa, but that there is evidence of such "slow green revolutions" going ahead in Africa (Eicher, Maredia, and Sithole-Niang 2006).

66. Pingali and Raney 2005; Ogodo 2007.

67. Alliance for a Green Revolution in Africa 2007.

68. In an interview in 2007, a leading South African scientist lamented that the Rockefeller Foundation, long a stalwart champion of biotechnology-based solutions, had also substantially withdrawn its support for the view that GMOs would be the silver bullet for African agriculture (interview with South African scientist, July 30, 2007).

69. See, for instance, Dickson 2007; NEPAD 2005; and Dhlamini 2006.

70. In October 2007, the Maize Breeders' Network, an organization of maize breeders, seed producers, and development specialists in southern and eastern Africa, called for faster regulatory approvals of locally bred conventional varieties that are adaptable to local conditions. This initiative was spearheaded by AGRA.

Conclusion

1. The same was true in India; see Herring and Roberts 2007; and Scoones 2006.

2. According to the United Nations' Food and Agriculture Organization, between 2005 and early 2008, world food prices rose 80 percent, putting the cost of food far beyond the reach of many of the world's poor. So acute was the crisis that urban workers and consumers around the world began rioting, at least one prime minister was deposed (in Haiti), and several countries placed restrictions on agricultural exports in order to keep enough food to feed their own people. In Japan, the United States, and many countries of Europe, all relatively rich, consumers also cut back on their consumption of expensive fruits, vegetables, and meat in favor of lower-priced food items.

3. Ron Herring (2007) details Indian farmers' and seed suppliers' illegal replication and use of Monsanto's patented seed technologies in a process he describes as "anarcho-capitalism." Brazilian farmers also planted GM seeds before GMOs were legally permitted in 2005, and only after Monsanto had created a postharvest system for collecting royalties was it able to charge for its intellectual property in Brazil and Argentina.

4. *Nature Biotechnology* 2008. "Heal, fuel, feed the world" was the slogan adopted by the Biotechnology Industry Organization's annual meeting in San Diego, California, in June 2008.

5. The 2008 final report of the International Assessment of Agricultural Knowledge, Science and Technology for Development (2008) clearly reveals these new directions in thinking. Similar trends are apparent in the work of the Alliance for a Green Revolution in Africa and in the recent funding decisions on agriculture made by the Gates Foundation.

6. Soule and Olszak 2004; Burstein 1999.

7. Giugni, McAdam, and Tilly 1999.

8. We take this phrase from Guigni et al.'s similarly titled 2008 book.

9. Tesh 2000.

10. Several authors have noted the importance of this distinction; see Herring and Roberts 2006; and Ruivenkamp 2005.

11. Of course, proponents of the technology loaded the term with meanings as well, hailing the gene revolution as a revolutionary advance in science and global agriculture. See, for instance, Borlaug 2000; and Paarlberg 2008.

12. Another source of rising costs was the technical difficulties scientists often encountered as they sought to develop particular technologies.

13. These persistent uncertainties were marked in a variety of locations. For instance, in April 2009, the U.S.-based Union of Concerned Scientists produced a report titled *Failure to Yield: Evaluating the Performance of Genetically Engineered Crops* (Gurian-Sherman 2009), which claimed that a broad review of peer-reviewed literature showed it was impossible to demonstrate systematic yield benefits to GM crops. In the same month, the Office of Technology Assessment in the German parliament produced a report titled *Transgenic Seeds in Developing Countries—Experience, Challenges, Perspectives,* which concluded, inter alia, that

because of weak data "the question of whether genetically modified plants can offer sustainable, regionally adapted options for differently developed agrarian systems in the medium- and long-term future cannot currently be answered in a substantiated way" (Office of Technology Assessment at the German Parliament 2009). The variability of yields among smallholders was also demonstrated by Gouse et al. (2009).

14. We do not seek to oppose this to the notion of rational action. Rather, we contend that all rationalities are profoundly (and invariably) cultural.

15. Evidence of this weakness is reflected in Europe's postmoratorium compromise on GMOs, which is based on a strategy of the peaceful coexistence of GM and non-GM crops. Because the European anti-biotech movement believes that no GMOs should be planted under any conditions, it has lost some of its sway with EU policymakers who want (or are feeling pressured) to take a more moderate position. The movement, however, has struggled to remain relevant and influential at the policy level.

Appendix

1. Martineau 2001.
2. Purdue 2000.
3. See especially Levidow and Carr 2007; Levidow 2003; and Carr 2000.

Abbott Laboratories. 1983. *Annual report 1982.* Abbott Park, Ill.

Abley, Mark. 2000. "The bio-battle of words." *The Gazette* (Montreal), January 29.

Aerni, Philipp. 2002. "Public attitudes towards agricultural biotechnology in South Africa." In *Final report, joint research project of CID (Harvard) and SALDRU (UCT),* 1–37. Cambridge, Mass.: Center for International Development, Harvard University; and Cape Town, South Africa: Southern Africa Labour and Development Research Unit, University of Cape Town.

Alliance for a Green Revolution in Africa. 2007. "As schools begin, unique African partnership announces launch of critical PhD program for crop breeding in Africa." Media release, Nairobi. http://www.agra-alliance.org/content/news/detail/647/.

———. N.d. "About" site. http://www.agra-alliance.org/section/about.

Amin, Ash, and N. J. Thrift. 2004. *The Blackwell cultural economy reader.* Malden, Mass.: Blackwell.

APDB. 2004. "Governing biotechnology in Africa: Towards consensus on key issues on biosafety." A living paper prepared for the 2nd session of the African Policy Dialogues on Biotechnology, September 2–4, Harare, Zimbabwe.

Ashton, G., G. Baker, M. Mayet, E. Pschorn-Strauss, and W. Stafford. 2004. "Objections to application for a permit for additional trials with insect resistant Bt Cry V genetically modified potatoes (*Solanum tuberosum L.* variety 'Spunta' G2 and G3), as applied for by Dr. G. Thompson, director of plant protection and biotechnology." South African Agricultural Research Council, Pretoria, May 24, 2003 (submitted June 28).

Bair, Jennifer. 2008. "Introduction: Global commodity chains: Genealogy and review." In *Frontiers of commodity chain research,* ed. Jennifer Bair, 1–34. Stanford, Calif.: Stanford University Press.

Barboza, David. 1999. "Biotech companies take on critics of gene-altered food." *New York Times,* November 11.

BBC News. 1998. "Iceland freezes out 'genetic' foods." BBC Online Network. London, March 18.

———. 1999a. "Sainsbury's phase out GM food." BBC Online Network. London, March 17.

———. 1999b. "UK: GM food firm rapped over adverts." BBC Online Network, London, August 11.

Beck, Ulrich. 1992. *Risk society: Towards a new modernity.* London: Sage Publications.

Beintema, Nienke, and Gert-Jan Stads. 2006. "Agricultural R&D in sub-Saharan Africa: An era of stagnation." In *Agricultural science and technology indicators (ASTI) initiative,* 44. Washington, D.C.: International Food Policy Research Institute.

Bellmann, Christophe, Graham Dutfield, and Ricardo Meléndez-Ortiz, eds. 2003. *Trading in knowledge: Development perspectives on TRIPS, trade, and sustainability.* London: Earthscan.

Bernauer, Thomas. 2003. *Genes, trade, and regulation: The seeds of conflict in food biotechnology.* Princeton, N.J.: Princeton University Press.

Biodiversity Action Thailand. 2000. "Thailand: Long march against GMOs." BIOTHAI, Nonthaburi, Thailand. http://wwwbiothai.org/cgi-bin/content/GMO/show.pl?0001, accessed September 12, 2000.

Bogève Declaration. 1988. "The Bogève Declaration: Towards a people-oriented biotechnology." *Development Dialogue* 1–2: 289–91. Dag Hammarskjöld Foundation. Bogève, France. Originally published 1987.

Bonneuil, Christophe, Pierre-Benoit Joly, and Claire Marris. 2008. "Disentrenching experiment: The construction of GM-crop field trials as a social problem." *Science, Technology and Human Values* 33: 201–28.

Borlaug, Norman E. 2000. "Ending world hunger: The promise of biotechnology and the threat of antiscience zealotry." *Plant Physiology* 124: 487–90.

Borlaug, Norman E., and Christopher Dowswell. 2001. "The unfinished Green Revolution—the future role of science and technology in feeding the developing world." Paper presented at the Seeds of Opportunity Conference, May 31–June 1. School of Oriental and African Studies, London.

Boudreaux, Karol. 2006. "Seeds of hope: Agricultural technologies and poverty alleviation in rural South Africa." *Mercatus Policy Series Policy Comment.* Mercatus Center at George Mason University, Arlington, Va.

Boyd, William. 2003. "Wonderful potencies? Deep structure and the problem of monopoly in agricultural biotechnology." In *Engineering trouble: Biotechnology and its discontents,* ed. Rachel Schurman and Dennis Takahashi Kelso, 24–62. Berkeley: University of California Press.

Brockett, Charles D. 1991. "The structure of political opportunities and peasant mobilization in Central America." *Comparative Politics* 23: 253–74.

Brunner, Eric. 1988. *Bovine somatotropin: A product in search of a market.* Report to the London Food Commission's BST Working Party. London: London Food Commission.

Bud, Robert. 1998. "Molecular biology and the long-term history of biotechnology." In *Private science: Biotechnology and the rise of the molecular sciences,* ed. Arnold Thackray, 3–19. Philadelphia: University of Pennsylvania Press.

Burstein, Paul. 1999. "Social movements and public policy." In *How social movements matter,* ed. Marco Giugni, Doug McAdam, and Charles Tilly, 3–21. Minneapolis: University of Minnesota Press.

Business and Industry Advisory Committee to the OECD. 1998. "An open and efficient global food system." Business and Industry Advisory Committee to the OECD, Paris. http://www.biac.org.

Butler, L. J. 1999. "The profitability of rBST on US dairy farms." *AgBioForum Online Journal* 2: 111–17.

Buttel, Frederick. 2003. "GMOs: The Achilles' heel of the globalization regime?" In *Engineering trouble: Biotechnology and its discontents,* ed. Rachel Schurman and Dennis Takahashi Kelso, 152–73. Berkeley: University of California Press.

Callon, Michel, ed. 1998. *Laws of the markets.* Oxford: Blackwell Publishers.

Carr, Susan. 2000. *EU safety regulation of genetically-modified crops: Summary of a ten-year country study.* Milton Keynes, UK: Open University.

Cavalieri, Liebe. 1976. "New strains of life—or death?" *New York Times Magazine,* August 22.

Charles, Daniel. 2001. *Lords of the harvest: Biotech, big money, and the future of food.* Cambridge, Mass.: Perseus Publishing.

Chinsembu, K., and T. Kambikambi. 2001. "Farmers' perceptions and expectations of genetic engineering in Zambia." *Biotechnology and Development Monitor* 47: 13–14.

Clapp, Jennifer. 2006. "Unplanned exposure to genetically modified organisms: Divergent responses in the global South." *Journal of Environment and Development* 15: 3–21.

Clark, Norman, John Mugabe, and James Smith. 2002. *Governing agricultural biotechnology in Africa: Building public confidence and capacity for policymaking.* Nairobi, Kenya: African Centre for Technology Studies.

Cock, Jacklyn. 2007. *The war against ourselves: Nature, power and justice.* Johannesburg, South Africa: Wits University Press.

Cohen, Deborah, and Carey Gillam. 2002. "Oregon GMO label defeat spells uphill battle ahead." Reuters News release, November 11.

Committee for Responsible Genetics. 1983. *GeneWatch: A newsletter of the Committee for Responsible Genetics.* Cambridge, Mass.

Cook, Louise. 1999. "Food or foul? Bio-battle is joined." *Business Day,* August 26 (Johannesburg, South Africa).

Crehan, Kate A. F. 1997. *The fractured community: Landscapes of power and gender in rural Zambia.* Berkeley: University of California Press.

Crespi, R. Stephen. 1999. "The biotechnology patent directive is approved at last!" *Trends in Biotechnology* 17: 139–42.

Culliton, Barbara J. 1976. "Recombinant DNA: Cambridge City Council votes moratorium." *Science* 193: 300–301.

Dawkins, Kristin. 1993. "Food self-reliance and the concept of subsidiarity: Alternative approaches to trade and international democracy." In *Hidden faces: Environment, development, justice: South Africa and the global context,* ed. David Hallowes, 103–17. Scottsville, South Africa: Earthlife Africa.

Dhlamini, Zephaniah. 2006. "The role of non-GM biotechnology in developing world agriculture." SciDevNet, February 1. http://www.scidev.net.

Dickinson, Susan. 1988. "Known for its good chemistry, DuPont goes multidisciplinary." *The Scientist* 2, no. 17 (September): 1.

Dickson, David. 2007. "Africa must create its own biotechnology agenda." SciDevNet, June 12. http://www.scidev.net.

DiMaggio, Paul, and Walter Powell. 1983. "The iron cage revisited: Institutional isomorphism and collective rationality in organizational fields." *American Sociological Review* 48: 147–60.

Doyle, Jack. 1985. *Altered harvest: Agriculture, genetics, and the fate of the world's food supply.* New York: Viking.

Du Gay, Paul, and Michael Pryke. 2002. *Cultural economy: Cultural analysis and commercial life.* London: Sage Publications.

DuPont. 1982. *Annual report 1981.* Wilmington, Del.

The Economist. 1998. "Food Fights." 347, no. 8072 (June 13): 79–80.

Egziabher, Tewolde Berhan G. 2000. "Biosafety negotiations—flashbacks." *Third World Resurgence,* no. 114–15: 24–26.

———. 2003. "Briefing: Facts to consider when receiving genetically engineered food aid." January 21. http://ngin.tripod.com/210103b.htm, accessed January 24, 2010.

Eichenwald, Kurt, Gina Kolata, and Melody Peterson. 2001. "Biotechnology food: From the lab to a debacle." *New York Times,* January 25.

Eicher, Carl, Karim Maredia, and Idah Sithole-Niang. 2006. "Crop biotechnology and the African farmer." *Food Policy* 31: 504–27.

Einwohner, Rachel L. 1999. "Practices, opportunity, and protest effectiveness: Illustrations from four animal rights campaigns." *Social Problems* 46: 169–86.

Einwohner, Rachel L., and J. William Spencer. 2005. "'That's how we do things here': Local culture and the construction of sweatshop and anti-sweatshop activism in two campus communities." *Sociological Inquiry* 75: 249–72.

Eisner, Marc Allen. 2000. *Regulatory politics in transition*. Baltimore, Md.: Johns Hopkins University Press.

Elias, Paul. 2002. "Biotech lobbyists' clout grows in Washington." Associated Press release, June 2.

Elliott, Howard, and Paul Perrault. 2006. "Zambia: Quiet crisis in African research and development." In *Agricultural R&D in the developing world: Too little, too late?* ed. Philip G. Pardey, Julian M. Alston, and Roley R. Piggott, 227–56. Washington, D.C.: International Food Policy Research Institute.

Elmer-DeWitt, Philip. 1987. "Tubers, berries and bugs." *Time*, May 11.

Emmott, Steve. 2001. "No patents on life: The incredible ten-year campaign against the European Patent Directive." In *Redesigning life? The worldwide challenge to genetic engineering*, ed. Brian Tokar, 361–72. London: Zed Books.

European Council. 1990. COUNCIL DIRECTIVE 90/220/EEC *on the deliberate release into the environment of genetically modified organisms*. Brussels: European Council.

Eyerman, Ron, and Andrew Jamison. 1991. *Social movements: A cognitive approach*. University Park: Pennsylvania State University Press.

Fernandez-Cornejo, Jorge. 2004. *U.S. seed industry in U.S. agriculture: An exploration of data and information on crop seed markets, regulation, industry structure, and research and development*. Agriculture Information Bulletin 786. U.S. Department of Agriculture, Economic Research Service, Washington, D.C. http://www.ers.usda.gov/publications/AIB786.

Fikes, Bradley. 1999. "Why San Diego has biotech." *San Diego Metropolitan*, April.

Fisher, Melissa S., and Greg Downey. 2006. *Frontiers of capital: Ethnographic reflections on the new economy*. Durham, N.C.: Duke University Press.

Fowler, Cary. 1978. *The Graham Center seed directory*. Wadesboro, N.C.: Rural Advancement Fund of the National Sharecroppers Fund.

Fowler, Cary, Eva Lachkovics, Pat Mooney, and Hope Shand. 1988. *The laws of life: Another development and the new biotechnologies*. Uppsala, Sweden: Dag Hammarskjöld Foundation.

Gamson, William A., and David S. Meyer. 1996. "Framing political opportunity." In *Comparative perspectives on social movements*, ed. Doug McAdam, John D. McCarthy, and Mayer Zald, 275–90. Cambridge: Cambridge University Press.

Ganz, Marshall. 2000. "Resources and resourcefulness: Strategic capacity in the unionization of California agriculture, 1959–1966." *American Journal of Sociology* 105: 1003–62.

Gardner, Den. 2001. "Council for Biotechnology Information: 'We need to reach the gatekeeper.'" *Agri Marketing* 39, no. 4 (April 1): 70–71.

Gaskell, George, Nick Allum, Martin Bauer, et al. 2000. "Biotechnology and the European public." *Nature Biotechnology* 18: 935–38.

Gaskell, George, Sally Stares, Agnes Allansdottir, et al. 2006. "Europeans and bio-technology in 2005: Patterns and trends: Final Report on Eurobarometer 64.3." In *A Report to the European Commission's Directorate General for Research*, 87. Brussels: Directorate General for Research, European Commission.

Gaskell, George, Paul Thompson, and Nick Allum. 2002. "Worlds apart? Public opinion in Europe and the USA." In *Biotechnology: The making of a global controversy,* ed. Martin Bauer and George Gaskell, 351–75. Cambridge: Cambridge University Press.

Gillis, Justin. 2000. "Frito-Lay's halfway measures banning GE corn freak out their competitors; new seed planted in genetic flap." *Washington Post,* February 6.

Giugni, Marco, Doug McAdam, and Charles Tilly. 1999. *How social movements matter.* Minneapolis: University of Minnesota Press.

Goodell, Rae. 1979. "Public involvement in the DNA controversy: The case of Cambridge, Massachusetts." *Science, Technology and Human Values* 4: 36–43.

Gottweis, Herbert. 1998a. *Governing molecules: The discursive politics of genetic engineering in Europe and the United States.* Cambridge, Mass.: MIT Press.

———. 1998b. "The political economy of British biotechnology." In *Private science: Biotechnology and the rise of the molecular sciences,* ed. Arnold Thackray, 105–30. Philadelphia: University of Pennsylvania.

Gouse, Marnus, Jenifer Piesse, Colin Thirtle, and Colin Poulton. 2009. "Assessing the performance of GM maize amongst smallholders in KwaZulu-Natal, South Africa." *AgBioForum* 12: 78–89.

GRAIN. 1993. "A decade in review." *Seedling* (a publication of Genetic Resources Action International) (February). http://www.grain.org/seedling/?id=377.

———. 1995. "The directive is dead." *Seedling* (March): 1–5. http://www.grain.org/seedling/?id=64.

Greenberg, Stanley. 1998. "Genetic engineering: Root of the matter." *The Guardian* (London), November 25.

Gupta, A., and R. Falkner. 2006. "The Cartagena Protocol on Biosafety and domestic implementation: Comparing Mexico, China and South Africa." Chatham House: *EEDP Briefing Paper 06/01,* March. London.

Gurian-Sherman, Doug. 2009. *Failure to yield: Evaluating the performance of genetically engineered crops.* Washington, D.C.: Union of Concerned Scientists.

Habermas, Jürgen. 1984. *The theory of communicative action.* Boston: Beacon Press.

Hart, Gillian Patricia. 2002. *Disabling globalization: Places of power in post-apartheid South Africa.* Berkeley: University of California Press.

Henson, Spencer, and James Northern. 1998. "Economic determinants of food safety controls in supply of retailer own-branded products in United Kingdom." *Agribusiness* 14: 113–26.

Herring, Ron. 2007. "Stealth seeds: Bioproperty, biosafety, biopolitics." *Journal of Development Studies* 43: 130–57.

Herring, Ron, and Kenneth Roberts. 2006. "Contentious politics: Science, social science, and social protest." Ithaca, N.Y.: Theme Proposal, Institute for Social Sciences, Cornell University. http://www.socialsciences.cornell.edu/0609/contentiouspolitics_iss_theme_proposal.pdf, accessed January 24, 2010.

Hindo, Brian, and Joshua Schneyer. 2007. "Monsanto: Winning the ground war: How the company turned the tide in the battle over genetically modified crops." *Business Week* 4063 (December 17): 34–41.

Ho, Karen. 2009. *Liquidated: An ethnography of Wall St.* Durham, N.C.: Duke University Press.

Hobbes, Jacklynne. 2009. "Seeds of the future." *Mail & Guardian* (Johannesburg, South Africa), January 29, 9.

Horner, Christopher. 2004. "Monsanto to realign research portfolio, development of Roundup Ready wheat deferred." Press release, May 10. Monsanto Company, St. Louis, Mo.

Howard, Ted, and Jeremy Rifkin. 1977. *Who should play God? The artificial creation of life and what it means for the future of the human race.* New York: Delacorte Press.

Ingersoll, Bruce. 1990. "Growth drug for milk cows draws more fire." *Wall Street Journal,* December 4.

International Assessment of Agricultural Knowledge, Science and Technology for Development (IAASTD). 2008. "Executive summary of the synthesis report." Johannesburg: IAASTD. Widely available online.

James, Clive. 2008. "Global status of commercialized biotech/GM crops: 2008." ISAAA brief no. 39. Edited by the International Service for the Acquisition of Agri-biotech Applications. Ithaca, N.Y.

Jasanoff, Sheila. 2005. *Designs on nature: Science and democracy in Europe and the United States.* Princeton, N.J.: Princeton University Press.

Jasper, James M. 1997. *The art of moral protest: Culture, biography, and creativity in social movements.* Chicago: University of Chicago Press.

Jelsoe, Erling, Jesper Lassen, Arne Thing Mortensen, Helle Fredericksen, and Mercy Wambui Kamara. 1998. "Denmark." In *Biotechnology in the public sphere: A European sourcebook,* ed. John Durant, Martin Bauer, and George Gaskell, 29–42. London: NSMI Trading Ltd.

Jelsoe, Erling, Jesper Lassen, Arne Thing Mortensen, and Mercy Wambui Kamara. 2001. "Denmark: The revival of national controversy over biotechnology." In *Biotechnology: 1996–2000, the years of controversy,* ed. George Gaskell and Martin Bauer, 157–71. London: Science Museum.

Jenson, Jane. 1995. "What's in a name? National movements and public discourse." In *Social movements and culture,* ed. Bert Klandermans and Hank Johnston, 107–26. Minneapolis, Minn.: University of Minnesota Press.

Joly, Pierre-Benoit, and Claire Marris. 2003. "La trajectoire d'un problème public: Une approche comparée du cas des OGM en France et aux Etats-Unis."

In *Risques collectifs et situations de crise: Apports de la recherche en sciences humaines et sociales,* ed. Claude Gilbert, 41–63. Paris: L'Harmattan.

Kameri-Mbote, Patricia. 2007. "Will Kenya's biosafety bill of 2005 ever become law?" SciDevNet, June 12. http://www.scidev.net.

Kaufman, Phil R. 2002. "Food retailing." *U.S. Food Marketing System, AER-11.* U.S. Department of Agriculture, Economic Research Service, Washington, D.C.

Kay, Lily. 1998. "Problematizing basic research in molecular biology." In *Private science: Biotechnology and the rise of the molecular sciences,* ed. Arnold Thackray, 20–38. Philadelphia: University of Pennsylvania Press.

Keck, Margaret E., and Kathryn Sikkink. 1998. *Activists beyond borders: Advocacy networks in international politics.* Ithaca, N.Y.: Cornell University Press.

Keeley, James, and Ian Scoones. 2003. "Contexts for regulation: GMOs in Zimbabwe." IDS Working Paper 190. Institute of Development Studies, Brighton, Sussex.

Kelemu, Segenut, George Mahuku, Martin Fregene, et al. 2003. "Harmonizing the agricultural biotechnology debate for the benefit of African farmers." *African Journal of Biotechnology* 2: 394–416.

Kelso, Dennis Doyle Takahashi. 2003. "Conclusion: Recreating democracy." In *Engineering trouble: Biotechnology and its discontents,* ed. Rachel Schurman and Dennis Doyle Takahashi Kelso, 239–54. Berkeley: University of California Press.

Kempf, Hervé. 2003. *La guerre secrète des OGM.* Paris: Seuil.

Kenney, Martin. 1986. *Biotechnology: The university-industrial complex.* New Haven, Conn.: Yale University Press.

Kitschelt, Herbert P. 1986. "Political opportunity structures and political protest: Anti-nuclear movements in four democracies." *British Journal of Political Science* 16: 57–95.

Klandermans, Bert, and Hank Johnston. 1995. *Social movements and culture.* Minneapolis, Minn.: University of Minnesota Press.

Kloppenburg, Jack Ralph. 2004. *First the seed: The political economy of plant biotechnology, 1492–2000.* 2nd ed. Madison: University of Wisconsin Press.

Kram, Jerry W. 2001. "Modified crops grow some attention." *Bismark Tribune* (Bismark, N.Dak.), February 7.

Kriesi, Hanspeter, Ruud Koopmans, Jan Willem Duyvendak, and Marco Giugni. 1992. "New social movements and political opportunities in western Europe." *European Journal of Political Research* 22: 219–44.

Krimsky, Sheldon. 1982. *Genetic alchemy: The social history of the recombinant DNA controversy.* Cambridge, Mass.: MIT Press.

———. 1988. "The release of genetically engineered organisms into the environment: The case of ice minus." In *Environmental hazards: Communicating risks as a social process,* ed. Sheldon Krimsky and Alonzo Plough. Dover, Mass.: Auburn House Publishing.

———. 1991. *Biotechnics and society: The rise of industrial genetics.* New York: Praeger.

———. 2003. *Science in the private interest: Has the lure of profits corrupted biomedical research?* Lanham, Md.: Rowman and Littlefield Publishers.

Kuyek, Devlin. 2002. "Intellectual property rights in African agriculture: Implications for small farmers." Briefing, August. Barcelona: Genetic Resources Action International (GRAIN). http://www.grain.org/docs/africa-ipr-2002-en.pdf.

Lagnado, Lucette. 1999. "Strained peace: Gerber baby food, grilled by Greenpeace, plans swift overhaul—gene-modified corn and soy will go, although firm feels sure they are safe—Heinz takes action too." *Wall Street Journal,* July 30.

Lambrecht, Bill. 1989. "'Hormone milk' stirs fear Monsanto working for consumer acceptance." *St. Louis Post-Dispatch,* January 29.

———. 1990a. "Expert questions safety of hormone ties between manufacturers, FDA too close, veterinarian says." *St. Louis Post-Dispatch,* January 14.

———. 1990b. "Spoiling for legal battle." *St. Louis Post-Dispatch,* January 28.

———. 1991. "Monsanto Co. says it will change its publicity campaign for a drug that raises cows' milk production." *St. Louis Post-Dispatch,* February 21.

———. 2001. *Dinner at the new gene café: How genetic engineering is changing what we eat, how we live, and the global politics of food.* New York: Thomas Dunne Books.

Lappé, Frances Moore, Joseph Collins, and Cary Fowler. 1977. *Food first: Beyond the myth of scarcity.* Boston: Houghton-Mifflin.

Leader Post. 2002. "NFU fears conflict in Monsanto's advisory panels." (Regina, Saskatchewan), November 13.

Lean, Geoffrey. 1999a. "GM foods—countdown to a consumer victory." *The Independent* (London), May 2.

———. 1999b. "GM foods—victory for grassroots action." *The Independent* (London), Devember 1.

Lenz, Kelly. 2003. "Growers, groups question company's intent." *Topeka Capital-Journal,* April 17.

Leskien, Dan. 1998. "The European patent directive on biotechnology." *Biotechnology and Development Monitor* 36: 16–19.

Levidow, Les. 1999. "Britain's biotechnology controversy: Elusive science, contested expertise." *New Genetics and Society* 18: 47–64.

———. 2000. "Pollution metaphors in the UK biotechnology controversy." *Science as Culture* 9: 325–51.

Levidow, Les, and Susan Carr. 2007. "GM crops on trial: Technological development as a real world experiment." *Futures* 39, no. 4: 408–31.

———. 2010. *GM food on trial: Testing European democracy.* New York: Routledge.

Levinson, Harry, and Stuart Rosenthal. 1984. *CEO: Corporate leadership in action.* New York: Basic Books.

Los Angeles Times. 1989. "5 food chains refuse milk of hormone-treated cows." August 24.

Losey, John, Linda S. Rayor, and M. E. Carter. 1999. "Transgenic pollen harms monarch larvae." *Nature* 399: 214.

Luker, Kristin. 1984. *Abortion and the politics of motherhood.* Berkeley: University of California Press.

Lydersen, Kari. 2002. "GM wheat portends disaster for Great Plains." AlterNet, September 9. http://www.alternet.org.

Magnan, André. 2007. "Strange bedfellows: Contentious coalitions and the politics of GM wheat." *Canadian Review of Sociology and Anthropology* 44: 289–316.

Magretta, Joan. 1997. "Growth through global sustainability—an interview with Monsanto's CEO, Robert B. Shapiro." *Harvard Business Review* 75, no. 1 (January–February): 78–88.

Mail & Guardian Online. 2006. "Parliament hears that legislation hampers research." Mail & Guardian Online, January 17. Capetown, South Africa.

Manda, Olga. 2003. "Controversy rages over 'GM' food aid." *African Renewal* 16: 5–6.

Markle, Gerald E., and Stanley S. Robin. 1985. "Biotechnology and the social reconstruction of molecular biotechnology." *Science, Technology, and Human Values* 10: 70–79.

Marris, Claire. 2000. "Swings and roundabouts: French public policy on agricultural GMOs since 1996." *Politeia* 60: 22–37.

Marris, Claire, Pierre-Benoit Joly, Stephanie Ronda, and Christophe Bonneuil. 2005. "How the French GM controversy led to the reciprocal emancipation of scientific expertise and policy making." *Science and Public Policy* 32: 301–8.

Martineau, Belinda. 2001. *First fruit: The creation of the Flavr Savr tomato and the birth of genetically engineered food.* New York: McGraw-Hill.

McAdam, Doug, John D. McCarthy, and Mayer N. Zald. 1996. "Introduction: Opportunities, mobilizing structures, and framing processes—toward a synthetic, comparative perspective on social movements." In *Comparative perspectives on social movements: Political opportunities, mobilizing structures, and cultural framings,* ed. Doug McAdam, John D. McCarthy, and Mayer N. Zald, 1–22. Cambridge: Cambridge University Press.

McAfee, Kathy. 2003. "Biotech battles: Plants, power and intellectual property in the new global governance regimes." In *Engineering trouble: Biotechnology and its discontents,* ed. Rachel Schurman and Dennis Doyle Takahashi Kelso, 174–94. Berkeley: University of California Press.

McCormick, Douglas. 1989. "Cave-in." *Bio/Technology* 7 (October): 981.

McMichael, Philip. 2000. *Development and social change: A global perspective.* Thousand Oaks, Calif.: Pine Forge Press.

McNair, Joel. 1990. "Dairy Board's BGH activity: Lawsuit." *Agri-View,* December 8.

Miller, Henry I. 1997a. *Biotechnology regulation: The unacceptable costs of excessive regulation*. London: Social Affairs Unit.

————. 1997b. *Policy controversy in biotechnology: An insider's view*. Austin, Tex.: R. G. Landes.

Millstone, Erik, Eric Brunner, and Ian White. 1994. "Commentary: Plagiarism or protecting public health?" *Nature* 371: 647–48.

Mnyulwa, D., and J. Mugwaga. 2003. "Agricultural biotechnology in Southern Africa: A regional synthesis." Working Paper no. 1, prepared for FANRPAN/IFPRI Regional Policy Dialogue on Biotechnology, Agriculture, and Food Security in Southern Africa. Johannesburg, April 23–25.

Monsanto Company. 1999. *1998 annual report*. St. Louis, Mo.

Mooney, Pat. 1979. *Seeds of the earth: A private or public resource?* Ottawa: Inter Pares for the Canadian Council for International Co-operation and the International Coalition for Development Action (London).

Moore, Kelly. 2008. *Disrupting science: Social movements, American scientists, and the politics of the military, 1945–1975*. Princeton, N.J.: Princeton University Press.

Moreton, Bethany. 2009. *To serve God and Wal-Mart: The making of Christian free enterprise*. Cambridge, Mass.: Harvard University Press.

National Academy of Sciences. 1977. *Research with recombinant DNA: An academy forum, March 7–9*. Washington, D.C.: National Academy of Sciences.

————. 1994. *Biographical memoirs,* vol. 63. Washington, D.C.: National Academy Press.

Nature. 2008. "A fruitless campaign." Editorial. *Nature* 456: 421–22.

Nature Biotechnology. 2008. "Join the dots: Pushing biotech as the 'solution' to the world's problems is doing more harm than good." *Nature Biotechnology* 26 (August): 837.

Nelkin, Dorothy, and Michael Pollack. 1981. *The atom besieged: Extraparliamentary dissent in France and Germany*. Cambridge, Mass.: MIT Press.

NEPAD. 2005. *Africa's science and technology consolidated plan of action*. http://www.nepadst.org/doclibrary/pdfs/doc27_082005.pdf.

Newman, Stuart. 2001. "Australian mouse study confirms CRG warning." *GeneWatch* 14, no. 2: 3.

Novartis. 1997. *Annual report, 1996*. Basel, Switzerland.

————. 2006. "About Novartis: Company History." Novartis, Basel, Switzerland. http://www.novartis.com/about-novartis/company-history/index.shtml.

Office of Technology Assessment at the German Parliament. 2009. *Transgenic seeds in developing countries—experience, challenges, perspectives*. Report no. 128, April 28, Berlin. http://www.tab.fzk.de/en/projekt/zusammenfassung/ab128.htm.

Ofir, Z. M. 1994. "Biotechnology in the new South Africa." *Biotechnology and Development Monitor* 20: 14–15.

Ogodo, O. 2007. "Kenyan maize variety resistant to boring pest." SciDevNet, October 18. http://www.scidev.net.

Oliver, Pamela E., and Hank Johnston. 2000. "What a good idea! Ideologies and frames in social movement research." *Mobilization* 4: 37–54.

Olson, R. Dennis. 2005. "Hard red spring wheat at a genetic crossroad: Rural prosperity or corporate hegemony?" In *Controversies in Science and Technology: From Maize to Menopause,* ed. Daniel Lee Kleinman, Abby J. Kinchy, and Jo Handelsman, 150–68. Madison: University of Wisconsin Press.

Paarlberg, Robert L. 2006. "Are genetically modified (GM) crops a commercial risk for Africa?" *International Journal of Technology and Globalization* 2: 81–92.

———. 2008. *Starved for science: How biotechnology is being kept out of Africa.* Cambridge, Mass.: Harvard University Press.

Palca, Joseph. 1986. "Another turn for ice-minus bugs." *Nature* 321: 190.

Pates, Mikkel. 2002. "'Deal made in dough': Monsanto, spring wheat bakers sign pact to perfect system to segregate wheats, especially genetically modified varieties." *Grand Forks Herald* (Grand Forks, N. Dak.), June 26.

Patterson, Lee Ann. 2000. "Biotechnology policy: Regulating risks and risking regulation." In *Policy-making in the European Union,* ed. Helen Wallace and William Wallace, 317–43. London: Oxford University Press.

Peoples Business Commission Amicus Brief. 1979. *Lutrelle F. Parker, acting commissioner of patents and trademarks, petitioner v. Malcolm E. Bergy, et al., Lutrelle F. Parker, acting commissioner of patents and trademarks, petitioner v. Ananda M. Chakrabarty,* no. 79–136, October term, 1979, December 13, 1979, "Brief on behalf of the Peoples Business Commission, amicus curiae." Leonard S. Rubenstein, Hirschkop & Grad, P.C., 108 N. Columbus Street, Alexandria, Va. 22312, for the Peoples Business Commission: Ted Howard, Dan Smith, Jeremy Rifkin, 1346 Connecticut Avenue, NW, no. 1010, Washington, D.C. 20036. 35 pp.

Perlas, Nicanor. 1994. *Overcoming illusions about biotechnology.* London and Atlantic Highlands, N.J.: Zed Books; and Penang, Malaysia: Third World Network.

———. 2003. Acceptance speech by Nicanor Perlas, at the Right Livelihoods Awards, December 8, Stockholm, Sweden. http://www.rightlivelihood.org/perlas_speech .html?&no_cache=1&sword_list[0]=nicanor&sword_list[1]=perlas.

Pingali, P., and T. Raney. 2005. "From the Green Revolution to the gene revolution: How will the poor fare?" ESA Working Paper no. 05-09, November. Food and Agriculture Organization.

Prudham, W. Scott. 2003. "Building a better tree: Genetic engineering and fiber farming in Oregon and Washington." In *Engineering trouble: Biotechnology and its discontents,* ed. Rachel Schurman and Dennis Takahashi Kelso, 63–83. Berkeley: University of California Press.

Purdue, Derrick A. 2000. *Anti-genetiX: The emergence of the anti-GM movement.* Aldershot, U.K.: Ashgate.

Rafferty, Tom. 2001. "Monsanto threatens to pull the plug on GMO research." Associated Press release. *Minot Daily News* (Bismarck, N.Dak.), March 10.

Reich, Robert. 2007. *Supercapitalism: The transformation of business, democracy, and everyday life.* New York: Knopf.

Richards, Bill. 1989. "Sour reception greets milk hormone." *Wall Street Journal,* September 15.

Roberts, Dan. 1999. "UK: Hazelwoods pledges to clear out GM foods." *Daily Telegraph* (London), June 16.

Rochon, Thomas R. 1998. *Culture moves: Ideas, activism, and changing values.* Princeton, N.J.: Princeton University Press.

Roosevelt, Margot. 2000. "Taking it to Main Street." *Time,* July 31.

Rootes, Chris. 2003. *Environmental protest in western Europe.* Oxford: Oxford University Press.

Rotman, David. 1993. "Agchem producers sow plans for a rich harvest." *Chemical Week* 153: 33–36.

Rucht, Dieter. 1990. "Campaigns, skirmishes, and battles: Anti-nuclear movements in the USA, France, and West Germany." *Industrial Crisis Quarterly* 4: 193–222.

Ruivenkamp, Guido. 2005. "Tailor-made biotechnologies: Between bio-power and sub-politics." *Tailoring Biotechnologies* 1: 11–33.

Schacter, Bernice Zeldin. 1999. *Issues and dilemmas of biotechnology: A reference guide.* Westport, Conn.: Greenwood Press.

Schmidta, Sascha, and Edwin Rühli. 2002. "Prior strategy processes as a key to understanding mega-mergers: The Novartis case." *European Management Journal* 20: 223–34.

Schneider, Keith. 1988. "Biotechnology's cash cow." *New York Times,* June 12.

———. 1990a. "Biotechnology enters political race." *New York Times,* April 21.

———. 1990b. "Wisconsin temporarily banning gene-engineered drug for cows." *New York Times,* April 28.

Schoenberger, Erica J. 1997. *The cultural crisis of the firm.* Cambridge, Mass.: Blackwell Publishers.

Schutz, Alfred, and Thomas Luckmann. 1973. *The structures of the life-world.* Evanston, Ill.: Northwestern University Press.

Schweiger, Thomas G. 2001. "Europe: Hostile lands for GMOs." In *Redesigning life? The worldwide challenge to genetic engineering,* ed. Brian Tokar, 361–72. London: Zed Books.

Scoones, Ian. 2005. "Contentious politics, contentious knowledges: Mobilising against GM crops in India, South Africa, and Brazil." *Institute for Development Studies Working Paper 256,* November.

————. 2006. *Science, agriculture and the politics of policy: The case of biotechnology in India.* New Delhi: Orient Longman.

Segarra, Alejandro, and Susan R. Fletcher. 2001. *Biosafety protocol for genetically modified organisms: Overview.* CRS Report no. RL30594, January 18. Washington, D.C.: Congressional Research Service, Library of Congress.

Seppa, Nathan. 1989. "BGH moratorium gains steam." *Wisconsin State Journal* (Madison), October 11.

————. 1990. "BGH moratorium survives." *Wisconsin State Journal* (Madison), March 23.

Shapiro, Robert. 1998. "Trade, feeding the world's people and sustainability: A cause for concern." In *The CEO series: Business leaders: Thought and action,* ed. Center for the Study of American Business, 7. St. Louis, Mo.: Washington University.

Sheingate, Adam. 2006. "Promotion versus precaution: The evolution of biotechnology policy in the United States." *British Journal of Political Science* 36: 243–68.

Shoemaker, Robbin, ed. 2001. *Economic issues in agricultural biotechnology.* Washington, D.C.: U.S. Department of Agriculture, Resource Economics Division, Economic Research Service.

Smith, Christian. 1996. *Resisting Reagan: The U.S. Central America peace movement.* Chicago: University of Chicago Press.

Snow, David A., and Robert D. Benford. 1988. "Ideology, frame resonance, and participant mobilization." In *From structure to action,* ed. Bert Klandermans et al., 197–218. Greenwich, Conn.: JAI Press.

————. "Master frames and cycle of protest." In *Frontiers in social movement theory,* ed. Aldon Morris and Carol Mueller, 133–55. New Haven, Conn.: Yale University Press.

Snow, David A., E. B. Rochford, S. K. Worden, and R. D. Benford. 1986. "Frame alignment processes, micromobilization, and movement participation." *American Sociological Review* 51: 464–81.

Soule, Sarah, and Susan Olszak. 2004. "When do movements matter? The politics of contingency and the Equal Rights Amendment." *American Sociological Review* 69: 473–97.

South African Government. 1997. "GMO Act (no. 15 of 1997) and GMO application process." Directorate Genetic Resources, Department of Agriculture, Pretoria.

————. 1998. "National Environmental Management Act (no. 107 of 1998)." Department of Environmental Affairs and Tourism, Pretoria, South Africa.

————. 2005. "Annual report on Genetically Modified Organisms Act." Department of Agriculture, Pretoria.

————. 2007. "National biotechnology audit 2007." Department of Science and Technology, Pretoria, South Africa.

Specter, Michael. 2000. "The Pharmageddon riddle." *New Yorker,* April 10, 58–71.

Szasz, Andrew. 1994. *Ecopopulism: Toxic waste and the movement for environmental justice.* Minneapolis: University of Minnesota Press.

Tangley, Laura. 1986. "Biotechnology on the farm." *BioScience* 36, no. 9 (October): 590–93.

Tarrow, Sidney G. 1989. *Democracy and disorder: Protest and politics in Italy, 1965–1975.* Oxford: Oxford University Press.

———. 2005. *The new transnational activism.* New York: Cambridge University Press.

Tesh, Sylvia Noble. 2000. *Uncertain hazards: Environmental activists and scientific proof.* Ithaca, N.Y.: Cornell University Press.

Thomas, Jim. 2001. "Princes, aliens, superheroes and snowballs: The playful world of the UK genetic resistance." In *Redesigning life? The worldwide challenge to genetic engineering,* ed. Brian Tokar, 357–50. London: Zed Books.

Thornley, Jennifer. 2001. "Got 'hormone-free' milk? Your state may have enough interest to let you know." *Indiana Law Journal* 76: 785–801.

Tiberghien, Yves, and Sean Starrs. 2004. "The EU as global trouble-maker in chief: A political analysis of EU regulations and EU global leadership in the field of genetically modified organisms." Paper presented at the Conference of Europeanists, Organized by the Council of European Studies (CES), Chicago, March 4.

Tjaronda, W. 2006. "What are the dangers of biotechnology?" New Era Publication Corp., Windhoek, Namibia, http://www.newera.com.na/page.php?id=4, accessed May 25, 2006.

Tokar, Brian. 2001. *Redesigning life? The worldwide challenge to genetic engineering.* London: Zed Books.

U.S. Congress. 1988. *New developments in biotechnology: Field-testing engineered organisms: Genetic and ecological issues.* Report OTA-BA-350, May. Washington, D.C.: Office of Technology Assessment.

USDA/ERS. 2002. "Outlook for U.S. agricultural trade." February 26, Edition AES-33. U.S. Department of Agriculture, Economic Research Service, Washington, D.C. http://usda.mannlib.cornell.edu/reports/erssor/trade/aes-bb/2002/aes33.pdf.

U.S. Wheat Associated Press Release. 2002. "European and American millers tell U.S. Wheat Associates Board to go slow on GM wheat." U.S. Wheat Associates, Washington, D.C., August 28. http://www.connectotel.com/gmfood/ap280802.txt.

van Roozendaal, G. 1995. "Organizing biotechnology research in Africa." *Biotechnology and Development Monitor* 23: 12–15.

Vos, Ellen. 2000. "EU food safety regulation in the aftermath of the BSE crisis." *Journal of Consumer Policy* 23: 227–55.

Wade, Nicholas. 2007. "Pursuing synthetic life, scientists transplant genome of bacteria." *New York Times,* June 29.

Wallach, Lori, and Patrick Woodall. 2004. *Whose trade organization? A comprehensive guide to the WTO*. New York: New Press.

Waugh, Paul. 1999. "British stores Tesco and Unilever ban genetically manipulated products." *The Independent* (London), April 28.

Weiss, Rick. 2002. "Despite famine, Zimbabwe balks at genetically altered corn." *International Herald Tribune* (Paris), August 1.

Wheale, Peter, and Ruth M. McNally. 1988. *Genetic engineering: Catastrophe or utopia?* New York: St. Martin's Press.

Wisner, Robert. 2003. GMO *spring wheat: Its potential short-term impacts on U.S. wheat exports markets prices*. Iowa State University, Department of Agricultural Economics, Ames, November 4. http://www.econ.iastate.edu/faculty/wisner/gmowheatreportMarch200311.pdf.

Woodward, B., J. Brink, and D. Berger. 1999. "Can agricultural biotechnology make a difference in Africa?" *AgBioForum* 2: 175–81.

World Bank. 2007. *World development report 2008: Agriculture for development*. Series report. Washington, D.C.: World Bank.

World Resources Institute. 2001. *The Monsanto Company: Quest for Sustainability*. Edited by Erik Simanis. Washington, D.C.: World Resources Institute.

Wright, Susan. 1994. *Molecular politics: Developing American and British regulatory policy for genetic engineering, 1972–1982*. Chicago: University of Chicago Press.

Yoon, Carol Kaesuk. 2008. "Genetic modification taints corn in Mexico." *New York Times*, October 2.

Zangerl, A. R., D. McKenna, C. L. Wraight, et al. 2001. "Effects of exposure to event 176 *Bacillus thuringiensis* corn pollen on monarch and black swallowtail caterpillars under field conditions." *Proceedings of the National Academy of Sciences (US) (PNAS)* 98, no. 21 (October 9): 11908–912.

Zawel, Stacey A. 2000. "Testimony of Stacey A. Zawel, vice president of scientific and regulatory policy at the Grocery Manufacturers Association, to the California Senate Agriculture Committee." March 28, Sacramento.

Zerbe, Noah. 2004. "Feeding the famine? American food aid and the GMO debate in southern Africa." *Food Policy* 29: 593–608.

Zimmerman, Bill, Len Radinsky, Mel Rothenberg, and Bart Meyers. 1972. "Towards a science for the people." http://socrates.berkeley.edu/~schwrtz/SftP/Towards.html.

economies, postindustrial, 202n1

economies, postwar: competitiveness in, 5; deregulation of, 5; neoliberalism in, 4, 5

economy, European: crisis of 1970s, 87

economy, global: biotechnology in, 13; civil activism in, 12; multilateral organizations in, 6; North–South relations in, 10–11; postwar governance of, 4–6; private property rights in, 7. *See also* globalization

economy, U.S.: deregulation of, 32; effect of regulation in, 124; role of biotechnology in, 126–27

Egypt: *Bt* cotton in, 176

Egziabher, Tewolde, 155, 220n7; on substantial equivalence, 222n31

Eicher, Carl, 224n65

Einwohner, Rachel, 201n28

Emmott, Steve, 211n23

environmental degradation: social movements concerning, 9, 11, 55

environmental justice, xxviii, 11; in South Africa, 166

Environmental Protection Agency (EPA): and *Pseudomonas syringae* controversy, 122, 123; role in biotechnology regulation, 124–26, 216n18

Environmental Rights Action (NGO), 220n10

eugenics, xii

Europe: biotechnology development in, 84, 86–88, 105; corporate culture in, 202n32; dynamics of contention in, xxvii; economic crisis in, 87; food culture of, 104; GMO moratorium in, x, 115–16, 168, 179; IPRs in, 84; quality-of-life movements in, 89. *See also* anti-biotechnology movement, European

European colonies, former: rejection of GMOs, 84

European Commission: biotechnology policies of, 88, 92; Directive 90/220, 98, 99, 112–13, 115, 212n31, 214n65; Directive 2001/18, 212n31; Directive on the Legal Protection of Biotechnological Inventions, 95–96, 212nn25–26; Directorates-General, 97, 212n28; Directorates-General for Science, Research, and Development, 98; Directorates-General for the Environment, 98–99, 212n33; GMO decisions by, 214n73; legislative role of, 211n21; "Life Patents Directive," 84; role in GMO approval, 112–13

European Parliament: and bST issue, 95; Green Group in, 92, 96; regulation of GMOs, 112, 113; rejection of genetic patenting, 96–97

European Patent Convention (1973), 211n23

European Union: anti-biotechnology movement in, 95–99; in Cartagena Protocol, 154; food regulation in, 84; GMO approval process in, 112–13; GMOs in, 88, 98; labeling requirements of, xxiii, 223n54; patent protection in, 95–96; regulation in, 97–99, 112–15; rejection of GMOs, 83–84; state–society relations in, 210n1; technology policy of, 85; trade with United States, 6

European Union Council: role in GMO approval, 112–13

Eyerman, Ron, 67, 209n40

farmers: biotechnology industry customers, xviii; in commodity chain, 109; corporate control of, 186; effect of Convention on Biological

Zambia: agricultural research and
development in, 222n30; during
food aid crisis, 157–58; view of
GMOs, 156

Zeneca (corporation): GM tomato
paste of, 99
Zimbabwe: beef exports of, 221n27;
biosafety policies in, 158; GM field
tests in, 222n46

(continued from page ii)

RACHEL SCHURMAN is associate professor of sociology and global studies at the University of Minnesota. She is coeditor of *Engineering Trouble: Biotechnology and Its Discontents.*

WILLIAM A. MUNRO is professor of political science and director of the international studies program at Illinois Wesleyan University. He is author of *The Moral Economy of the State: Conservation, Community Development, and State-Making in Zimbabwe.*